Estimating Benefits of Regional Solutions for Water and Wastewater Service

Awwa Research Foundation

About the Awwa Research Foundation

The Awwa Research Foundation (AwwaRF) is a member-supported, international, nonprofit organization that sponsors research to enable water utilities, public health agencies, and other professionals to provide safe and affordable drinking water to consumers.

The Foundation's mission is to advance the science of water to improve the quality of life. To achieve this mission, the Foundation sponsors studies on all aspects of drinking water, including supply and resources, treatment, monitoring and analysis, distribution, management, and health effects. Funding for research is provided primarily by subscription payments from approximately 1,000 utilities, consulting firms, and manufacturers in North America and abroad. Additional funding comes from collaborative partnerships with other national and international organizations, allowing for resources to be leveraged, expertise to be shared, and broad-based knowledge to be developed and disseminated. Government funding serves as a third source of research dollars.

From its headquarters in Denver, Colorado, the Foundation's staff directs and supports the efforts of more than 800 volunteers who serve on the board of trustees and various committees. These volunteers represent many facets of the water industry, and contribute their expertise to select and monitor research studies that benefit the entire drinking water community.

The results of research are disseminated through a number of channels, including reports, the Web site, conferences, and periodicals.

For subscribers, the Foundation serves as a cooperative program in which water suppliers unite to pool their resources. By applying Foundation research findings, these water suppliers can save substantial costs and stay on the leading edge of drinking water science and technology. Since its inception, AwwaRF has supplied the water community with more than $300 million in applied research.

More information about the Foundation and how to become a subscriber is available on the Web at **www.awwarf.org.**

Estimating Benefits of Regional Solutions for Water and Wastewater Service

Prepared by:
John Cromwell
Stratus Consulting Inc.
1920 L St. NW, Suite 420, Washington, DC 20036
and

Scott Rubin
Bloomsburg, PA

Jointly sponsored by:
Awwa Research Foundation
6666 West Quincy Avenue, Denver, CO 80235-3098
and

Lehigh County Authority
1053 Spruce, St. , P.O. Box 3348, Allentown, PA 18106

Co-Funding Provided by:
Renew Lehigh Valley
1337 E Fifth St., Bethlehem, PA 18015

Published by:

Awwa Research Foundation

Distributed by:

American Water Works Association

IWA Publishing

DISCLAIMER

ISBN 978-1-84339-629-1

CONTENTS

TABLES vii

FIGURES ix

FOREWORD xi

ACKNOWLEDGMENTS xiii

EXECUTIVE SUMMARY xv

CHAPTER 1: INTRODUCTION 1
- Project Objectives 2
- The Lehigh Valley is a Good Example for Benefits Analysis 4
- Testing a Method for Quantifying and Documenting Benefits 4
- Summary of the Project Approach 5
  - Task 1—Data Collection 5
  - Task 2—Define and Model Status Quo Scenarios 6
  - Task 3—Define and Model Regional Scenarios 6
  - Task 4—Comparison of Status Quo and Regional Scenarios 7
- Guide to the Project Report 7

CHAPTER 2: BASELINE ASSESSMENT 9
- Population Growth and Land Development 9
- Governance and Provision of Public Services 10
- Financial Risks of Water and Wastewater Capacity Management 15
- Financial Challenges of Replacing Aging Pipes 18
- Financial Implications of Utility Workforce Attrition 20

CHAPTER 3: DATA AND METHODOLOGY 23
- Overview 23
- Data Collection 23
- Status Quo Modeling 24
- Modeling Risk 27
- Regional Scenario Modeling 27

CHAPTER 4: RESULTS 31
- Water 31
  - Status Quo Projected to 2020 31
  - Consolidation Into Three Sub-Regional Utilities 32
  - Consolidation Into One Regional Utility 33
  - Testing the Comparability of Model Results 35

Wastewater 36
Status Quo Projected to 2020 36
Consolidation Into Sub-Regional Utilities 37
Consolidation Into One Regional Utility 39
Overall Impact on Water and Wastewater Costs 39

CHAPTER 5: FINDINGS AND CONCLUSIONS 41
Findings 41
Conclusions 42

CHAPTER 6: RECOMMENDATIONS 43

CHAPTER 7: RESEARCH IMPLICATIONS 45

APPENDIX A: DETAILED MODELING INPUTS AND ASSUMPTIONS 47

APPENDIX B: INSTRUCTIONS FOR USING THE WATER MODEL 49

REFERENCES 53

ABBREVIATIONS 55

TABLES

2.1 U.S. Census population data from 1990 and 2000 11

FIGURES

2.1 Lehigh Valley publicly owned community water supply systems 13

2.2 Lehigh Valley central and community sewer systems..14

2.3 Lehigh Valley growth forecasts compared to actual growth 15

2.4 Water and wastewater debt as a share of total city debt 17

2.5 Historical population growth in the Lehigh Valley 19

4.1 Total residential water costs in 2005 and 2020 status quo projected................................31

4.2 Water utility size and typical household cost—present and status quo projected to 2020 32

4.3 Total residential water costs in 2005 and 2020 status quo and sub-regional projected..33

4.4 Water utility size and typical household cost—present, status quo, and sub-regional projected to 2020..34

4.5 Total residential water costs in 2005 and 2020 status quo, sub-regional, and regional projected...34

4.6 Total residential wastewater costs in 2005 and 2020 status quo projected.......................36

4.7 Wastewater utility size and typical household cost—present and status quo projected to 2020..37

4.8 Total residential wastewater costs in 2005 and 2020 status quo and sub-regional projected..38

4.9 Wastewater utility size and typical household cost—present, status quo, and sub-regional projected to 2020..38

4.10 Total residential wastewater costs in 2005 and 2020 status quo, sub-regional, and regional projected...39

4.11 Total residential water and wastewater costs in 2005 and 2020 status quo, sub-regional, and regional projected...40

FOREWORD

The Awwa Research Foundation is a nonprofit corporation that is dedicated to the implementation of a research effort to help utilities respond to regulatory requirements and traditional high-priority concerns of the industry. The research agenda is developed through a process of consultation with subscribers and drinking water professionals. Under the umbrella of a Strategic Research Plan, the Research Advisory Council prioritizes the suggested projects based upon current and future needs, applicability, and past work; the recommendations are forwarded to the Board of Trustees for final selection. The foundation also sponsors research projects through the unsolicited proposal process; the Collaborative Research, Research Applications, and Tailored Collaboration programs; and various joint research efforts with organizations such as the U.S. Environmental Protection Agency, the U.S. Bureau of Reclamation, and the Association of California Water Agencies.

This publication is a result of one of these sponsored studies, and it is hoped that its findings will be applied in communities throughout the world. The following report serves not only as a means of communicating the results of the water industry's centralized research program but also as a tool to enlist the further support of the nonmember utilities and individuals.

Projects are managed closely from their inception to the final report by the foundation's staff and large cadre of volunteers who willingly contribute their time and expertise. The foundation serves a planning and management function and awards contracts to other institutions such as water utilities, universities, and engineering films. The funding for this research effort comes primarily from the Subscription Program, through which water utilities subscribe to the and consultants and manufacturers subscribe bbased on their annual billings. The program offers a cost-effective and fair method for funding research in the public interest.

A broad spectrum of water supply issues is addressed by the foundation's research agenda: resources, treatment and operations, distribution and storage, water quality and analysis, toxicology, economics, and management. The ultimate purpose of the coordinated effort is to assist water suppliers to provide the highest possible quality of water economically and reliably. The true benefits are realized when the results are implemented at the utility level. The foundation's trustees are pleased to offer this publication as a contribution toward that end.

David E. Rager
Chair, Board of Trustees
Awwa Research Foundation

Robert C. Renner, P.E.
Executive Director
Awwa Research Foundation

ACKNOWLEDGMENTS

This report was funded primarily by Renew Lehigh Valley via participating utilities in the Lehigh Valley of Pennsylvania, with matching funding from the Awwa Research Foundation (AwwaRF). Utility participants included Easton/Easton Suburban, Bethlehem, Allentown and the Lehigh County Authority. The project was managed and administered by Ms. Susan Turnquist under the AwwaRF tailored collaboration research program.

The researchers extend their great appreciation to AwwaRF, and to Susan Turnquist for management and encouragement. We also wish to thank the AwwaRF project advisory committee (PAC), which provided peer review comments, guidance, support, and critiques of this project. The PAC helped shape and improve this effort. The PAC members were:

Peter Binney, City of Aurora Utilities, Aurora, Colo.
Barry Gullet, Charlotte-Mecklenburg Utility, Charlotte, N.C.
David Kay, Cornell University, Ithaca, N.Y.
Dean Moss, Beaufort-Jasper Water & Sewer Authority, Okatie, S.C.

Special thanks are due to the Lehigh Valley Planning Commission (LVPC) for providing the research team with access to their extensive research and data resources. The LVPC also provided meeting space to the project team on numerous occasions, enabling extensive interactions with stakeholders representing the interests of more than half of the utilities in the Lehigh Valley.

Additional meeting accommodations were provided to the project by Muhlenberg College. In addition, several stakeholder meetings were facilitated by Dr. Christopher Borick, Associate Professor in the Political Science Department. The research team is most indebted to Kevin Conrad and Sharol Lilly Weaver of the Community Action Committee of the Lehigh Valley for their work in arranging meetings with stakeholders and encouraging participation in the project. Of course, the deepest debts are to those representing utilities throughout the Lehigh Valley who took precious time from their business to participate in the project and keep an open mind about possibilities for broader regional collaboration. It is the hope that this report will provide some repayment and offer the prospect of more that is worth discussing.

Thank you all very much.

John Cromwell
Stratus Consulting Inc.
Washington, D.C. (April 2008)
Scott Rubin
Bloomsburg, Pa.

EXECUTIVE SUMMARY

The Lehigh Valley in Eastern Pennsylvania is home to more than 600,000 people. The Lehigh Valley includes three significant urban centers—the Cities of Allentown, Bethlehem, and Easton—several smaller older communities (usually organized as boroughs), and numerous outlying suburban and rural areas (organized as townships). The Lehigh Valley borders New Jersey to the east, the Pocono Mountains to the north, the Philadelphia suburbs to the south, and the farmland of Berks County to the west.

The Lehigh Valley is one of the highest growth areas of Pennsylvania. Between 1990 and 2000, the Lehigh Valley's population grew by about 7.6%. From 2000 through 2005, it grew by another 6.7%. And from 2005 through 2020, it is projected that the population will increase by another 15% or more. In addition, and contrary to the trend in most of the Valley, western Lehigh County has seen a dramatic increase in large-volume customers in the form of beverage producers.

This growth has created, and will continue to cause, significant strains on land use policies and the provision of water and wastewater services. The Lehigh Valley's cities and boroughs are stagnant or even losing population, while the surrounding (formerly rural) areas are experiencing dramatic growth. Indeed, some townships have seen their population grow by 50% or more in just 10 years. Furthermore, the collective tax base of the region's three cities increased by only 0.5% between 1991 and 2006. In contrast, the collective tax base of the region's townships increased by 51.9% during the same time period.

Most water and wastewater services are provided in the Lehigh Valley by publicly owned utilities. Most of these are owned and controlled by individual municipalities. A number of them provide wholesale services to adjacent jurisdictions through various forms of agreements. There are two multi-jurisdictional municipal authorities that provide water service (and to a lesser extent wastewater service), one serving the primarily suburban portions of Lehigh County outside of Allentown, and one serving the City of Easton and its surrounding suburbs. In addition, the Cities of Allentown and Bethlehem own large wastewater treatment facilities that serve as regional treatment facilities for many of the surrounding municipalities. Allentown's and Bethlehem's water treatment facilities also provide wholesale (both cities) and retail (Bethlehem) water service to some surrounding communities. For the most part, though, water and wastewater services are provided on a disaggregated basis by 40 separate municipal utilities.

The purpose of this study is to assess the potential economic benefits from enhanced regional collaboration among these publicly owned water and wastewater utilities. Achieving economic benefits is important not only to save money for the region's consumers, but also to improve the utilities' economic viability and their ability to meet public health and environmental requirements.

RESEARCH OBJECTIVES AND APPROACH

This research project was undertaken to demonstrate a method of quantifying the potential benefits of regional collaboration between utilities. The methods employed are straightforward. After obtaining financial reports and conducting interviews with about half of the utilities in the Lehigh Valley, spreadsheet models were developed to project the finances of typical small, medium, and large utilities. Separate models were developed for water utilities and wastewater utilities. In addition, a separate set of models was developed for utilities that obtain treatment capacity as a wholesale customer of another utility while owning and operating only their distribution or collections systems. These are generic models, that is, they are based on the data collected from specific systems, but the data was averaged together for systems in the same size range to produce an amalgamated "typical" system, but not a replica of any specific system.

The use of such financial models to project future conditions consists of applying varying assumptions regarding the future expenses of the business. The key output of each model is the cost per household that is needed to keep the utility financially viable. Capital cost and risks entailed in treatment capacity, costs of infrastructure replacement, and costs related to human capital are some of the critical variables that will determine the future course of water and wastewater rates in the Lehigh Valley. The results are driven mainly by these trends.

The timeframe of financial analysis also is an essential consideration. Despite the long-term nature of the capital investments (20 years or more) and attendant financial risks involved in providing water and wastewater service, many utilities proceed incrementally and have only a short-term outlook on the future (five years or less). In contrast, the authors' methodology forecasts the financial condition of the model utilities in 2020, from a starting point grounded in 2005 annual financial reports. It is uncertain exactly when the financial risks will catch up with Lehigh Valley utilities over this period of time. But it is very certain that these cost drivers will exert their influence at some time during this period. Thus, the results projected for 2020 show where utilities will end up without trying to forecast the exact path they may traverse in getting there.

The authors first project results for a "status quo"—or, "no structural change" scenario—in which utilities are assumed to continue to operate in the same fashion as today with no change in the level or nature of regional collaboration. This scenario exposes the model utilities to the projected financial and operational risks of the future assuming they will take no advance actions (beyond what they might do today) to prepare for them. The authors next adjust assumptions to project results for various scenarios of enhanced regional collaboration. To simplify the analysis, the authors examine the alternative of consolidation into a single entity responsible for providing service to the entire Lehigh Valley and another alternative of creating three sub-regional consolidated utilities. The three sub-regions roughly correspond to areas surrounding the three major cities in the Lehigh Valley: Allentown, Bethlehem, and Easton. A few outlying areas in the northern portion of the Lehigh Valley would remain independent under this sub-regional scenario.

These consolidation cases show how enhanced collaboration helps to manage the financial risk of the future. The difference in cost—and therefore, in water and wastewater rates—between the status quo scenario and the consolidation scenarios in 2020 is taken as the authors' measure of the economic benefit of enhanced collaboration within the region. Organizational consolidation is only one form of enhanced collaboration but these benefit estimates can also serve as an indication of how much benefit, albeit lower, might be obtainable through "softer" forms of enhanced collaboration, involving less significant forms of institutional change. It also should be noted that the authors did not model any type of physical interconnection or consolidation, beyond what exists today. Significant additional savings might be achievable through consolidating treatment plants or interconnecting pipe networks, but the authors' results do not reflect any of these potential savings.

RESEARCH OUTPUTS

Under the status quo scenario, the average household in the Lehigh Valley, paying about $220 per year for water service in 2005, will pay about $510 per year in 2020. The same average household pays about $230 per year for wastewater service in 2005, which will escalate to about $480 per year in 2020 under the status quo scenario. So, the total water and wastewater bill for the average household is projected to nearly double from $450 per year in 2005 to $990 per year in 2020. Computing results for the average household masks considerable variation in the results. At the extremes, some small communities could see a tripling or quadrupling of the household cost by the year 2020 with combined water and wastewater bills approaching $1,800 per year or more in some communities.

Results for the sub-regional approach to consolidation that would create three large water and wastewater supply entities in the Lehigh Valley (surrounding Allentown, Bethlehem, and Easton) indicate that the combined water and wastewater bill for the average household would be about $820 per year, divided equally between water and wastewater. By comparison to the status quo scenario, this represents an average household savings of more than $170 per year. Regionwide, this amounts to a total savings of $41 million per year by 2020. It is important to stress that these savings would accrue not just once, but every year.

Results for the total consolidation scenario where all water and wastewater suppliers would be owned and managed by a single regional entity indicate that 2020 water bills for the average household would be about $380 per year and wastewater bills would be about $350 per year. By comparison to the status quo scenario, this represents an average household savings of $260 per year. Regionwide, this amounts to a total savings of $56 million per year by 2020. Again, these savings would accrue not just once, but every year.

CONCLUSIONS

Water and wastewater costs are likely to increase significantly faster than the rate of inflation in the Lehigh Valley during the next dozen years. The magnitude of these cost increases are not inevitable. There is currently a window of opportunity during which public water and wastewater utilities in the Lehigh Valley could save tens of millions of dollars each year through enhanced regional collaboration. Dramatic savings can be achieved through improved planning, financial management, risk reduction, facilities planning, infrastructure management, and workforce management. If action is not taken soon, however, many of these opportunities will be lost.

There are opportunities to manage and mitigate the risks and long-term consequences associated with population growth. But those opportunities will no longer exist after the growth has occurred. Given the level of projected population growth within the next dozen years, the delay of more than a few years will result in a significant lost opportunity.

During the next 15 years, Allentown will need to upgrade its wastewater treatment plant, and Lehigh County Authority will need new wastewater capacity. Collaboration between these neighboring systems may be mutually beneficial due to economies of scale in the design and construction of new or upgraded facilities. Similarly, Bethlehem will need to substantially upgrade its water and/or wastewater treatment plants within the next few years. If regional plans are not developed soon, the opportunities to enhance the utilization of these regional facilities may be missed.

Addressing the aging of the water and wastewater pipe infrastructure is a long-range challenge. The key to meeting that challenge is to undertake a comprehensive data collection, analysis, and planning process. But that process must begin within the next few years, or the costs of dealing with the failure to act will soon overwhelm utility budgets and other resources.

The looming retirement of the "baby boom" generation creates a unique opportunity to optimize the utility workforce throughout the Lehigh Valley without anyone losing their job. But failing to design and implement these plans within the next few years will miss this opportunity.

In summary, a number of factors have converged to make this an ideal time to consider an enhanced level of consolidation and regional cooperation in the Lehigh Valley. The authors' estimate that such a regional effort could save the average household as much as $260 per year in 2020—amounting to $56 million in annual savings for the region as a whole.

RECOMMENDATIONS

This research project was devised in part to follow up on recommendations of prior AwwaRF research (Raucher et al. 2006) which had called for a new paradigm to evaluate regional collaboration. It was observed that the traditional approaches of weighing pros and cons provide an unbalanced view because potential benefits are not made more tangible. When the potential economic benefits are vaguely defined, there is not enough motivation to counter the inertia of the status quo. When everything seems fine in the short-term, there is a tendency to not fix what does not appear broken. Thus, the scales used in the comparison of the pros and cons are not balanced. The new paradigm is called "benefits and issues." The intent is to first quantify the potential benefits of regional collaboration and then convene a process to discuss the benefit estimates and to identify and discuss the issues that would need to be addressed in order to devise a means of capturing those benefits.

This study of 40 utilities in the Lehigh Valley of Pennsylvania has demonstrated that there are substantial potential benefits to be won from enhanced regional collaboration. It is therefore a logical next step to convene a process to discuss the means of capturing these benefits and to identify potential issues that must be resolved. The Pennsylvania Municipality Authorities Act provides a very flexible framework within which new institutions could be created at either the sub-regional (e.g., Easton, Easton Suburban) or regional level to consolidate utility assets and functions to deliver more sustainable and affordable water and wastewater services to the benefit of all jurisdictions.

Regional collaboration raises a range of legitimate issues that are of concern to everyone. This is where the enormous range of flexibility provided within the Pennsylvania Municipality Authorities Act is extremely valuable. It is important to stress that, under the Act, the starting point for creation of a *new* authority, at either the sub-regional or regional level, would be a completely blank "clean sheet of paper." The parties wishing to design a new collaboration could design it from scratch to suit their needs and ameliorate any concerns they have. They can design features to prevent them from losing control through the charter and through their representation on the Board. They can address concerns over service continuity, quality, and cost control by stipulating performance requirements and guarantees. They can provide hands-on management of workforce transition issues. They can work through the due diligence details of asset transfers to achieve consolidation in a transparent process that will provide a clear basis for making arrangements as equitable as possible for all concerned. In short, every issue that needs to be resolved in order to capture the benefits identified in this report can be addressed on a "clean sheet of paper." No one has to sign on until they are fully satisfied that their concerns have been addressed. With benefits now identified, the authors' believe it is timely to convene a process to discuss the issues involved in moving forward to capture those benefits.

There is an opportunity to meet the challenges of the future with alternative approaches to regional collaboration that can increase the Lehigh Valley's ability to respond to change. But many of the benefits computed here would result from preparing for change before it occurs, or responding to change as it is happening, rather than responding after the fact. In order to capture these benefits, it is necessary to implement broader regional collaboration relatively soon. The most challenging cost influences that are bearing down on the utilities in the Lehigh Valley are visible today and may cause many individual utilities to make irreversible commitments that will negate the longer-term opportunity.

Lost opportunity is not the only motivation for early action to promote broader collaboration. It is clear that in the current configuration, individual jurisdictions are so limited in their ability to cope with change and mitigate risk that many of their available responses result in the acceptance of extraordinary levels of financial risk for public enterprises. At some point, this is not sustainable. The authors project that some utilities in the Lehigh Valley could see a tripling or even quadrupling of water and wastewater bills by 2020 unless they take action soon to manage the inherent risks of the business. It is also probable that limitations on bonding capacity and revenue generation could be constrained by affordability concerns or debt service coverage requirements. This could force utilities to forego needed programs for maintenance, rehabilitation, and/or capital improvement. Bonding or affordability constraints also would increase the financial and operational risks to a utility, ultimately affecting its ability to provide the level of high-quality service that its customers expect.

CHAPTER 1
INTRODUCTION

Throughout much of North America, changes in the levels and patterns of economic activity and land development are altering both the physical and socioeconomic landscape within which water and wastewater agencies operate. This suite of changes—including declining or slow-growing urban centers and traditional industrial activity clusters, contrasted with increasing suburbanization of once rural areas—creates new challenges for utilities. The levels and locations of service demands often correspond poorly to the types and levels of infrastructure and institutions that emerged over the prior century. This can create excess capacity and stranded assets in areas of economic and population decline (e.g., older urban and industrial centers). It also can spawn high demands for new capacity and new infrastructure in more dispersed and under-served areas (e.g., small towns being transformed by rapid but low density population growth).

When water and wastewater utilities in these settings operate in isolation, they face many challenges that could be met more advantageously with some form of regional cooperation. Low-growth and declining urban and industrial entities need to support a now over-sized portfolio of capital and water resource assets with a shrinking customer base. Concurrently, neighboring smaller entities face the challenge of meeting increased capital and water resource demands—as well as more stringent regulatory requirements. Regional solutions can benefit not just the impacted utilities, but also the greater regional community. The entire region benefits by having more efficient, cost-effective, and sustainable infrastructure arrangements that can help re-invigorate and sustain economic activity and provide higher quality service.

Despite the apparent logic and benefits of regional solutions, there is tremendous inertia in the status quo that must be overcome in order to induce change. Recent Awwa Research Foundation (AwwaRF) research on this issue (Raucher et al. 2006) pointed out that the benefits of regional approaches are often discussed in only the broadest and vaguest conceptual terms. There is a need to quantify those benefits. Absent a more concrete assessment of benefits, regional approaches will remain much discussed but infrequently implemented. Most success stories have resulted when overt crisis conditions related to water scarcity, regulatory compliance pressure, natural disaster, or other compelling forces have inspired regional solutions out of necessity. But where there is no pending crisis at hand, regional approaches are less likely to progress to the point of making an objective assessment of the potential benefits.

This "tailored collaboration" project was conceived to fill this gap by providing a fully developed practical example of how to quantify economic benefits of regional solutions. This research project developed empirical insights into the types and levels of benefit that regional approaches can provide to the Lehigh Valley region of Pennsylvania. The results provide a documented example and a replicable approach to help guide other regions and entities as they consider the potential value of regional approaches in their own areas.

Local funding for the project was provided through a grant from Renew Lehigh Valley and 50% matching funding was provided by the AwwaRF. Renew Lehigh Valley is a campaign to transform the Lehigh Valley into a more vibrant, economically competitive community with one regional identity, strong cities and boroughs that retain and attract young people, and a commitment to preserving open space. It is a diverse coalition of business and civic leaders, education experts, elected representatives, government officials, environmentalists, developers, and realtors formed to advance a policy agenda to reverse the troubling trends of the past several decades. Renew Lehigh Valley embraces a "smart growth" agenda that is exploring a variety of policy solutions that might include consolidations and cooperative efforts among municipalities or authorities.

The project was also supported by in-kind contributions of four participating utilities within the Lehigh Valley: Lehigh County Authority, City of Allentown, City of Bethlehem, and City of Easton/Easton Suburban. Significant local support was also provided by staff of the Community Action Committee of the Lehigh Valley. In addition, as part of the "tailored collaboration" process, AwwaRF provided contract management with an experienced project manager and peer review by a panel of utility executives and academic experts from other parts of the country having experience in regional collaboration.

PROJECT OBJECTIVES

The experience of Lehigh Valley is similar to that of many other regions in metropolitan and suburban areas across the United States (U.S.). Regional approaches have taken hold in many places—up to a point. Everywhere results are mixed. It is a fair question to ask, "If the economics are so compelling, why is there no stampede?" The reason for this is a blind spot. This project is intended to remove the blind spot by demonstrating a practical method of quantifying the comparative economic benefits of collaborative regional approaches.

Businesses must continually change and transform themselves in response to changes taking place in the physical, technologic, economic, demographic, and cultural environment in which they are operating. To do anything less would cause them to lose customers to more agile competitors and threaten the sustainability of the enterprise. That is a major reason why the value of a share of stock is based more heavily on the assessment of future performance than past performance. Simply, the ability to cope with change is key to success. And, successful businesses always operate with a forward-looking business plan to anticipate change. They may still experience turbulence along the way when market forces shift unexpectedly, but good businesses are prepared to sustain themselves because they have given some forethought to the possibilities.

Water and wastewater utilities in the Lehigh Valley are almost all owned by local governments, but they are nonetheless operating very challenging businesses—having to continually raise capital and balance revenues and expenses. For most of them, the ability to adapt to changing circumstances is greatly restricted, however, by static institutional structures and fixed jurisdictional boundaries. Forty publicly owned utilities provide water and/or wastewater services in Lehigh and Northampton counties. These entities were created over the course of the last century and, for the most part, are still anchored in assumptions that more closely reflect the business conditions that existed at the time of their creation than those of today. The next dozen years will produce significant changes in the cost of providing service that signal an ever greater rate of departure from the past. But public entities embedded within larger general government structures do not typically engage in the type of forward-looking business planning that is warranted in such circumstances.

The normal perspective in governmental entities is short-term; and, in the short-term, things can generally look okay. Financial reports reviewed in this study showed all the utilities to be solvent in 2005. Even when painful adversities are encountered, they tend to be viewed as bumps in the road without concern for what lies beyond the next bend. With no long-term view of how things are changing, there is no indication of a need for change. This is the blind spot and it can be a recipe for disaster. A sustainable utility business is one that knows what is coming, understands its options for adaptation, and then implements a strategy to adapt. It is essential to ask questions about the longer term. This project develops and describes a spreadsheet modeling approach that allows assessment of the long-term finances of a utility to address the blind spot.

The methodology forecasts the financial condition of typical Lehigh Valley utilities in 2020, from a starting point based on 2005 annual financial reports. The analysis shows where utilities could end up in 2020 without trying to forecast the exact path they may traverse in getting there. In the model—as in the real world—it is uncertain exactly when the cost drivers and financial risks the authors project to be advancing over this period will be experienced by Lehigh Valley utilities. But it is believed that they will exert their influence by 2020. Thus, the output characterizes only the endpoint.

In the comparison, the authors first project results in 2020 for a "status quo" scenario—or, "do nothing" scenario—in which the authors assume that utilities continue to operate in the same institutional structures as today with no change in the level or nature of regional collaboration. The authors next adjust assumptions to project results for various scenarios of enhanced regional collaboration. To simplify the analysis, the authors examine the alternative of consolidation into a single entity responsible for providing service to the entire Lehigh Valley and another alternative of creating three sub-regional consolidated utilities.

These regional consolidation alternatives show how enhanced collaboration helps to manage the financial risk of the future. The difference in household cost between the status quo scenario and the consolidation scenarios in 2020 is the authors' measure of the economic benefit of enhanced collaboration within the region. Organizational consolidation is only one form of enhanced collaboration but these benefit estimates can also serve as an indication of how much benefit, albeit lower, might be obtainable through "softer" forms of enhanced collaboration, involving less significant forms of institutional change. For simplicity, the model does not include any type of physical interconnection of facilities beyond what exists today.

The authors' spreadsheet model is provided with this report. It is pre-populated with the generic financial profiles of different types and sizes of utilities that the authors developed in this study. However, it allows the user to modify the inputs based on their own financial profile as reflected in their own financial report. It then allows the user to do as the authors have done and make varying assumptions to project the status quo and alternatives for 15 to 20 years into the future.

The heart of this approach is to focus first and foremost on defining and quantifying potential benefits. With that part of the picture in better focus as a guidepost, it is then a more straightforward matter to identify and address the range of issues that naturally arise in considering expanded regional collaboration. The process of working through the issues to finally assess feasibility is simply easier to embrace once there is a more concrete understanding of the mutual benefits that are being sought. It makes it easier to calibrate one's interest level. Vagueness about benefits is believed to be a major reason why regional initiatives stall out. When organizational change, political action, or legislative change is required, it is hard to overcome the inertia of the status quo without something substantive to say about economic benefits. This project fills that gap.

THE LEHIGH VALLEY IS A GOOD EXAMPLE FOR BENEFITS ANALYSIS

The Lehigh Valley provides a good road test opportunity for this new paradigm. A 2003 study by the Brookings Institution (Katz et al. 2003) squarely identified the fragmentation of government services such as utility services as an underlying constraint to economic prosperity in the Lehigh Valley. With this explicit benefits analysis, the intention is to consider a broad range of collaborative strategies encompassing all of the public water and wastewater utilities in the Lehigh Valley. This encompasses forty existing local government organizations involved in providing water and/or wastewater services.

Given the total project budget of $150,000, development of such analyses encompassing so many individual entities was a significant challenge that deserves further discussion. The new paradigm developed in the prior AwwaRF project encourages identification and documentation of benefits as the first step, and identification and resolution of issues as the second step. This project has focused on demonstrating the methodology for accomplishing the first step. While the results will be useful to stakeholders in the Lehigh Valley in undertaking the second step of issue identification and resolution, it is envisioned to occur after this project. This partitioning helped to define a clear and tractable objective for the project, while also producing the outputs needed to promote further discussions in the Lehigh Valley.

TESTING A METHOD FOR QUANTIFYING AND DOCUMENTING BENEFITS

Whereas some water and wastewater regional solutions have been brought about successfully in Lehigh Valley and elsewhere using economic and financial analyses to document the benefits, such analyses often perish in the files without ever attaining much notice. The literature is particularly sparse in recording this analytical process. This project intends to fill this void by providing a worked-out example. However, even with this narrowed focus, it still required a lot of analysis to produce results across 40 utilities within the proposed project budget. This highlights another objective of the project: how to devise a streamlined and efficient method of assessing benefits in order to produce such an analysis within practical constraints.

There is a tendency to go overboard in conducting such economic and financial analyses because many times people have difficulty differentiating between a *pro forma* level of analysis appropriate for a feasibility assessment, and a *due diligence* level of analysis necessary to execute a transaction. This project has adopted a strategic, top-down approach to obtaining an acceptable *pro forma* analysis that is sufficient to assess and document potential benefits from regional solutions while not getting lost in the details. Getting a sense of the general magnitude or scale of benefits is the essential element required initially to enable participants to calibrate their motivations. Moreover, developing too much detail too early can be counter-productive. The initial appraisal of the scale of potential benefits is best kept as simple as possible to make it understandable and verifiable in terms of common sense. This can make it more likely to be an acceptable guide for further discussions. Hence, the simplicity and transparency emphasized in the approach are as important a part of the demonstration as anything else.

In summary, the project objective was to develop and demonstrate an efficient and transparent methodology for developing an initial appraisal of economic benefits of regional solutions. This was accomplished using a worked-out example that will help convey the technique at a practical level and also providing the results needed to move the deliberative process to the next phase in the Lehigh Valley.

SUMMARY OF THE PROJECT APPROACH

The overall approach constituted a phased process that started with data collection and utility interviews to understand the status quo situation of utilities in the Lehigh Valley. The primary focus was on their current economic and financial performance. Based on this assessment, the project team next developed a baseline projection of the economic and financial implications of projecting the status quo for 15 years into the future—the status quo scenario. The status quo—or no action—scenario assumes that the same organizational entities face the changing business risks of the future with the same operational capabilities that exist today. Importantly, a year-by-year analysis was not performed. Rather, the focus was placed on what things will look like 15 years in the future if there are no structural or organizational changes.

This baseline analysis addresses the identified blind spot. The adequacy of operating entities is often asserted in terms of current conditions when in fact conditions are changing. Establishing a baseline that factors these changes into the evaluation of the status quo provides a sounder basis for projecting the future business performance of the organization—in the same manner that is applied in developing business plans for many other types of enterprise.

The baseline analysis also includes an important element that is often overlooked in this type of analysis or projection: risk. For a water system, risk can be manifested in many ways, including source water contamination, contract disputes with major suppliers, the loss of a major customer, natural disasters such as floods and droughts, changing regulatory requirements, among others. The baseline modeling analysis includes an explicit risk component (assuming that a water system could face a major disruption about once every 30 years) that results in important cost changes for some utilities.

In the final phase of the analysis, several conceivable strategies for enhanced regional collaboration are articulated and analyzed. The analysis focused on the same date in the future, essentially asking the question: "How would costs in 2020 be affected if structural or organizational change occurs in the interim?" The economic and financial results of these scenario analyses were compared to the status quo scenario. These differentials constitute the quantification of benefits that is the intended output of the analysis.

The specific tasks undertaken in the project methodology are summarized below.

Task 1—Data Collection

The project team began by inviting representatives of all 40 water and wastewater utilities in the Lehigh Valley to attend a project overview meeting that featured a presentation describing the project plan and discussion. Following the meeting, the utilities were invited to schedule individual interviews with the project team to share an understanding of the status quo condition of their facilities and operations. They were asked to bring a standard set of data, comprised of the elements necessary to develop a simple financial model of each utility. These included the following:

- Organizational structure and decision making authority
- Annual financial report
- Details of outstanding debt
- Annual operating budget; revenue requirements
- Existing collaborative arrangements, water purchases, etc.

- Staffing and payroll and equipment owned and/or shared
- Age, retirement plans, and succession plans for key management personnel
- Facilities owned, capacities and flows, age and condition (if available)
- Current rates (including hook-up fees and all charges)
- 5 or 10 year capital improvement plan
- Longer-term capital needs
- Population and service growth projections

About half of the 40 utilities accepted the invitation to meet and participated in the data collection process for the study.

Task 2—Define and Model Status Quo Scenarios

The project team used the data provided to develop spreadsheet models with which to simulate their financial future of the status quo from 2005 (the most recent financial report data) until 2020. In order to simplify the analysis and preserve some level of anonymity for the participating utilities, similar utilities were grouped into categories. Separate models for each generic category were developed to reflect the representative operating and financial profile of the actual utility data in each category. In this way, the models are well grounded in the empirical input data provided and representative of the category, but do not reflect any specific utility. This approach also makes it feasible to generalize about non-participating utilities.

Task 3—Define and Model Regional Scenarios

Once the initial status quo models were developed, the project team invited all the utilities to a meeting to present findings for the status quo scenario. The status quo scenario projected the financial outcomes if the present utility configurations are maintained to meet the operating and cost challenges of the future.

At the same meeting, candidate regional collaboration options in the Lehigh Valley were also put forward for discussion and further consideration. The focus of the regional options was on large scale regional collaboration options wherein assets and/or operating responsibilities are transferred to a larger regional entity in order to better respond to changing cost drivers and operating demands. Other forms of regional collaboration such as purchased water transactions were not considered in this effort. This limitation to larger-scale options helped to contain the range of options to a manageable number. In fact, the boundlessness of the range of options for regional collaboration was identified in earlier AwwaRF research as an impediment to progress on regional solutions because it creates an information overload that can be too great to assimilate. It is better to start with something specific. The very broadest initiatives may offer the greatest long-term benefits for the region, so it makes a good choice as a place to begin.

The broadest option considered was the formation of a single regional entity. Segmentation of an overall entity into several large sub-regional service areas based on differences in the cost of service or geography was also considered as an alternative. These options were analyzed by adapting the spreadsheet models to simulate the finances of larger regional or sub-regional entities in facing the future.

Task 4—Comparison of Status Quo and Regional Scenarios

A final meeting with utility representatives was then convened to present the results for the candidate regional scenarios, while also to solicit comment and feedback.

The comparisons that emerged from the models are summarized in terms of the household cost for water service and wastewater service. Comparing results shows how these costs will change over time in the regional scenarios versus the status quo scenario. This was also converted to a total cost savings to the region over the planning period between 2005 and 2020.

GUIDE TO THE PROJECT REPORT

This report provides documentation of the above-described analytical steps in a straightforward manner. Chapter 2 provides a baseline characterization of the status quo configuration of water and wastewater service provision in the Lehigh Valley and of the major cost drivers that are already in evidence. Chapter 3 summarizes the data collection process that was undertaken to more fully comprehend the status quo and the cost drivers. It also includes a description of the spreadsheet modeling methods that were employed to project financial outcomes from 2005 to 2020 under status quo versus regional collaboration scenarios. This is further supported by a more detailed description of the spreadsheets contained in appendix A. Copies of the models are also included on a compact disk (CD) with the report, with a guide to their use in appendix B. Chapter 4 documents the results of the analysis. Chapter 5 summarizes the project findings and conclusions. Chapter 6 presents the recommendations for the next steps in the Lehigh Valley. Chapter 7 discusses implications for further research.

CHAPTER 2
BASELINE ASSESSMENT

The Lehigh Valley of Pennsylvania is located in the central eastern portion of Pennsylvania within a 300-mile radius of the major East Coast cities of New York, Philadelphia, Baltimore, Washington, D.C., and Boston. This urban and rapidly suburbanizing region is also identified as the Allentown/Bethlehem/Easton Metropolitan Statistical Area, encompassing Lehigh County, Northampton County, and Carbon County. Carbon County is predominantly rural, however, and is not yet a major part of these urban/suburban trends. The analysis in this report considers the Lehigh Valley to be the 725-square mile area that lies within Lehigh and Northampton counties.

POPULATION GROWTH AND LAND DEVELOPMENT

Over the last few decades, the Lehigh Valley has been transformed by sweeping economic and demographic trends. The region entered a period of declining economic activity following the decline of the manufacturing sector that had been the mainstay of its economy throughout much of the Twentieth Century. Within the past 20 years, the region's proximity to major urban centers such as New York and Philadelphia has spurred population growth and land development along with an accompanying shift in emphasis from manufacturing employment to service sector employment.

The Lehigh Valley is now one of the highest growth areas of Pennsylvania. Between 1990 and 2000, the Lehigh Valley's population grew by about 7.6%. From 2000 through 2005, it grew by another 6.7%. And from 2005 through 2020, it is projected that the population will increase by another 15% or more. Between 2000 and 2020, the Lehigh Valley has been on a path to grow from just under 600,000 people to nearly 700,000 people—essentially adding another city the size of Allentown. Although recent macro economic trends affecting lending have produced a pause in this trend, the local market forces driving it remain in place.

The new growth and economic activity has brought new problems that have affected the region very unevenly. Population growth has been occurring almost exclusively in the outlying suburban regions. In these areas, township units of local government have been faced with a risky fiscal challenge of taking on rapid expansion of services to meet residential and service sector development. The accompanying land development is exacting a high price in terms of lost landscape values associated with the displaced rural/agricultural setting. Whereas the rate of population growth in the suburbs is equivalent to adding the same number of people as the City of Allentown every 20 years, the new development is consuming the same amount of land area as the City of Allentown every five years. At the same time, the older municipalities have failed to recover the population or investment that was lost with the decline of the old manufacturing base. Populations in Allentown, Bethlehem, Easton, and many of the boroughs that comprise their inner suburbs have been stable or declining. Housing vacancy rates in these older areas are double those in the outer suburbs. This is leading to a cycle of declining property values and depressed tax revenues. The new prospect for a casino development in Bethlehem has been the only major investment initiative of a scale that begins to rival what was lost.

GOVERNANCE AND PROVISION OF PUBLIC SERVICES

Current population and development trends constitute a threat to the ability of local governments to sustain the delivery of public services such as water and wastewater. Older cities and boroughs have increasing elderly and lower-income populations at a time when major reinvestments are needed in many of the region's larger water and wastewater facilities that were constructed during better economic times. Suburban townships are taking on financial risks to build new water and sewer capacity to meet the demands of new development. Although serving a younger, higher-income population and buffeted by connection fees to reduce the risks for the moment, these townships will eventually reach build-out. Some may get a preview of the impact of this if development slows sufficiently in the wake of the subprime mortgage crisis to affect cash flows that were planned to support existing commitments to capacity expansion.

These relationships between economic development and local government processes in Pennsylvania have been the subject of considerable study by The Brookings Institution (Katz et al. 2003). Pennsylvania has one of the most highly fragmented systems of local government in the U.S. The Lehigh Valley is typical, having 62 individual units of municipal government within Lehigh and Northampton counties: 3 cities, 27 boroughs, and 32 townships. That is about 10 units of government per 100,000 people, compared to a national average of 6 per 100,000. Such fragmentation is identified by the Brookings Institution as one of the chief mechanisms that promotes the false economy of suburban sprawl and undermines broader potential for regional economic development. Table 2.1 presents U.S. Census population data from 1990 and 2000 compiled by the Lehigh Valley Planning Commission that illustrates the extreme contrast between fast-growing jurisdictions and stable or declining jurisdictions within the Lehigh Valley. One of the contributors to the Brookings Institution studies presented the following analysis comparing the economic effects of fragmentation in the Lehigh Valley to the Charlotte-Mecklenburg metropolitan area in North Carolina, which has one of the most unified governmental structures (Rusk 2003).

The City of Charlotte and the Allentown-Bethlehem urbanized area are comparable in area and population. Comparing property valuation among states is very unreliable since formal assessment methodology and informal practices vary so widely. However, in 2002, the market value of property in Charlotte was about $54 billion; Charlotte's general obligation bonds carried a blue-chip, Aaa credit rating (one of only 21 municipalities in the country with such a rating). Charlotte executes almost $500 million a year in capital improvement projects. Charlotte can finance by itself what it needs for itself and for its region.

Compare that situation with the Lehigh Valley. Adjusting assessed valuation to market value, Lehigh and Northampton counties had a combined property tax base of about $30 billion in 2002. However, the largest of the "little boxes" are the City of Allentown, the market value of whose property was about $3.5 billion, and the City of Bethlehem, about $2.7 billion. Allentown and Bethlehem go to the bond market with only A3 and Baa2 bond ratings—six and eight ratings steps below Charlotte, respectively. The only local governments that could hope to match resources with the City of Charlotte would be the two county governments themselves, acting jointly and as agents for 62 constituent municipalities. Their broader, deeper tax bases are reflected in their quite respectable bond ratings: Lehigh County (Aa3) and Northampton County (Aa2).

Table 2.1
U.S. Census population data from 1990 and 2000

Municipality	2000 population	1990 population	Absolute change	Percent change
Lehigh County				
Alburtis	2,117	1,415	702	49.6
Allentown	106,632	105,301	1,331	1.3
Bethlehem (part)	19,029	18,866	163	0.9
Catasauqua	6,588	6,662	–74	–1.1
Coopersburg	2,582	2,599	–17	–0.7
Coplay	3,387	3,267	120	3.7
Emmaus	11,313	11,157	156	1.4
Fountain Hill	4,614	4,637	–23	–0.5
Hanover Twp.	1,913	2,033	–120	–5.9
Heidelberg Twp.	3,279	3,250	29	0.9
Lower Macungie Twp.	19,220	16,871	2,349	13.9
Lower Milford Twp.	3,617	3,269	348	10.6
Lowhill Twp.	1,869	1,602	267	16.7
Lynn Twp.	3,849	3,220	629	19.5
Macungie	3,039	2,597	442	17.0
North Whitehall Twp.	14,731	10,827	3,904	36.1
Salisbury Twp.	13,498	13,401	97	0.7
Slatington	4,434	4,678	–244	–5.2
South Whitehall Twp.	18,028	18,261	–233	–1.3
Upper Macungie Twp.	13,895	8,757	5,138	58.7
Upper Milford Twp.	6,889	6,304	585	9.3
Upper Saucon Twp.	11,939	9,775	2,164	22.1
Washington Twp.	6,588	6,356	232	3.7
Weisenberg Twp.	4,144	3,246	898	27.7
Whitehall Twp.	24,896	22,779	2,117	9.3
Lehigh County totals	312,090	291,130	20,960	7.2
Northampton County				
Allen Twp.	2,630	2,626	4	0.2
Bangor	5,319	5,383	–64	–1.2
Bath	2,678	2,358	320	13.6
Bethlehem (part)	52,300	52,561	–261	–0.5
Bethlehem Twp.	21,171	16,425	4,746	28.9
Bushkill Twp.	6,982	5,512	1,470	26.7
Chapman	234	254	–20	–7.9
East Allen Twp.	4,903	4,572	331	7.2
East Bangor	979	1,006	–27	–2.7
Easton	26,263	26,276	–13	0.0
Forks Twp.	8,419	5,923	2,496	42.1
Freemansburg	1,897	1,946	–49	–2.5
Glendon	367	391	–24	–6.1
Hanover Twp.	9,563	7,176	2,387	33.3
Hellertown	5,606	5,662	–56	–1.0
Lehigh Twp.	9,728	9,296	432	4.6
Lower Mt. Bethel Twp.	3,228	3,187	41	1.3
Lower Nazareth Twp.	5,259	4,483	776	17.3

(continued)

Table 2.1 (continued)

Municipality	2000 population	1990 population	Absolute change	Percent change
Northampton County (continued)				
Lower Saucon Twp.	9,884	8,448	1,436	17.0
Moore Twp.	8,673	8,418	255	3.0
Nazareth	6,023	5,713	310	5.4
North Catasauqua	2,814	2,867	–53	–1.8
Northampton	9,405	8,717	688	7.9
Palmer Twp.	16,809	14,965	1,844	12.3
Pen Argyl	3,615	3,492	123	3.5
Plainfield Twp.	5,668	5,444	224	4.1
Portland	579	516	63	12.2
Roseto	1,653	1,555	98	6.3
Stockerton	687	641	46	7.2
Tatamy	930	873	57	6.5
Upper Mt. Bethel Twp.	6,063	5,476	587	10.7
Upper Nazareth Twp.	4,426	3,413	1,013	29.7
Walnutport	2,043	2,055	–12	–0.6
Washington Twp.	4,152	3,759	393	10.5
West Easton	1,152	1,163	–11	–0.9
Williams Twp.	4,470	3,982	488	12.3
Wilson	7,682	7,830	–148	–1.9
Wind Gap	2,812	2,741	71	2.6
Northampton County totals	267,066	247,105	19,961	8.1
Lehigh Valley totals	579,156	538,235	40,921	7.6

Charlotte and Mecklenburg County might as well be formally consolidated. The city houses 78% of the county's population and has negotiated treaties with the county's other five municipalities so that ultimately all of Mecklenburg County will have been annexed out. City and county government occupy a joint government center, provide many unified services, have a single planning department and jointly appointed planning commission, and even have a common Web page (www.charmeck.nc.us).

To add to this comparison, Charlotte-Mecklenburg has consolidated their water and wastewater utilities into a single entity to serve the entire region of 750,000 people (http://www.charmeck.org/Departments/Utilities/Home.htm). Moreover, utility finances are totally independent of city finances. Water and sewer rates fund a $208 million operating budget. Capital investment of $78 million per year is supported with stand-alone utility revenue bonds that carry a AAA rating. The utility employed 818 people in 2007, averaging 3.5 per 1,000 water accounts. The average annual household bill for water and wastewater service in 2007 was $555. Although this cost is comparable to the amount charged by various utilities in the Lehigh Valley today, there is a key difference in sustainability.

The Charlotte-Mecklenburg utility has a long record of earning top national awards for quality service, financial performance, and environmental performance. It has the organizational maturity and wherewithal to face the future with confidence that this level of business performance is sustainable as reflected in its bond rating. Whether large or small, urban or rural, retail or wholesale, the 40 separate publicly owned utilities providing water and/or wastewater services in the Lehigh Valley all face continuously mounting financial challenges as they confront growing needs to replace aging infrastructure and provide new infrastructure to serve robust and yet fickle development trends. Each of the 40 entities is a separate business unit, facing these risks alone—most often as a part of a general purpose local government that is facing other competing challenges and priorities as well.

Figure 2.1 presents a map of the publicly owned community water supply systems in the Lehigh Valley developed by the Lehigh Valley Planning Commission. This map does not include a few small investor-owned water systems in the region, or Pennsylvania-American Water Company which has a larger investor-owned water system in portions of Northampton County. It also excludes small private well systems serving individual developments as well as "non-community" systems serving isolated stand-alone facilities. Figure 2.2 presents the Lehigh Valley Planning Commission's map of "central" and community sewer systems in the Lehigh Valley. The "central" systems are small isolated facilities serving schools or other small remote sites. The community sewer systems are the larger systems owned and operated by local government entities.

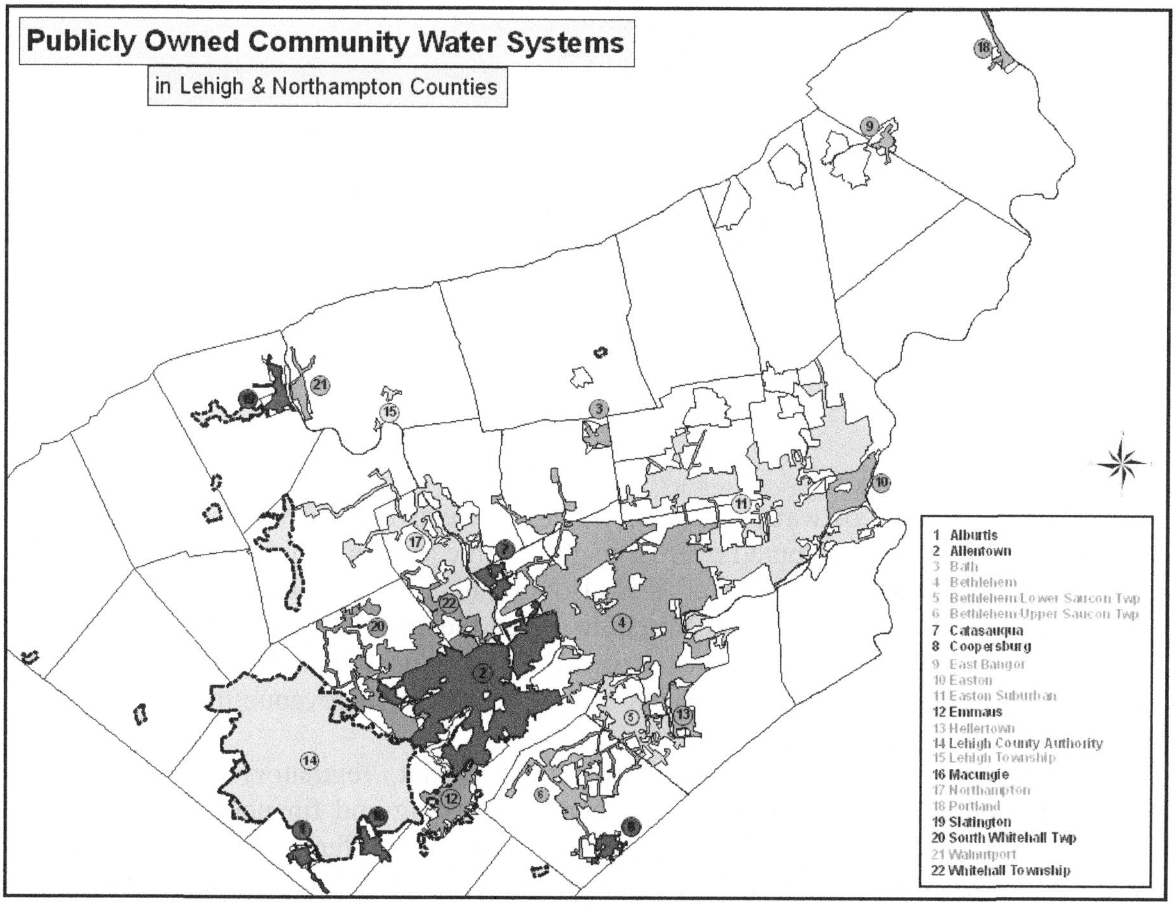

Figure 2.1 Lehigh Valley publicly owned community water supply systems

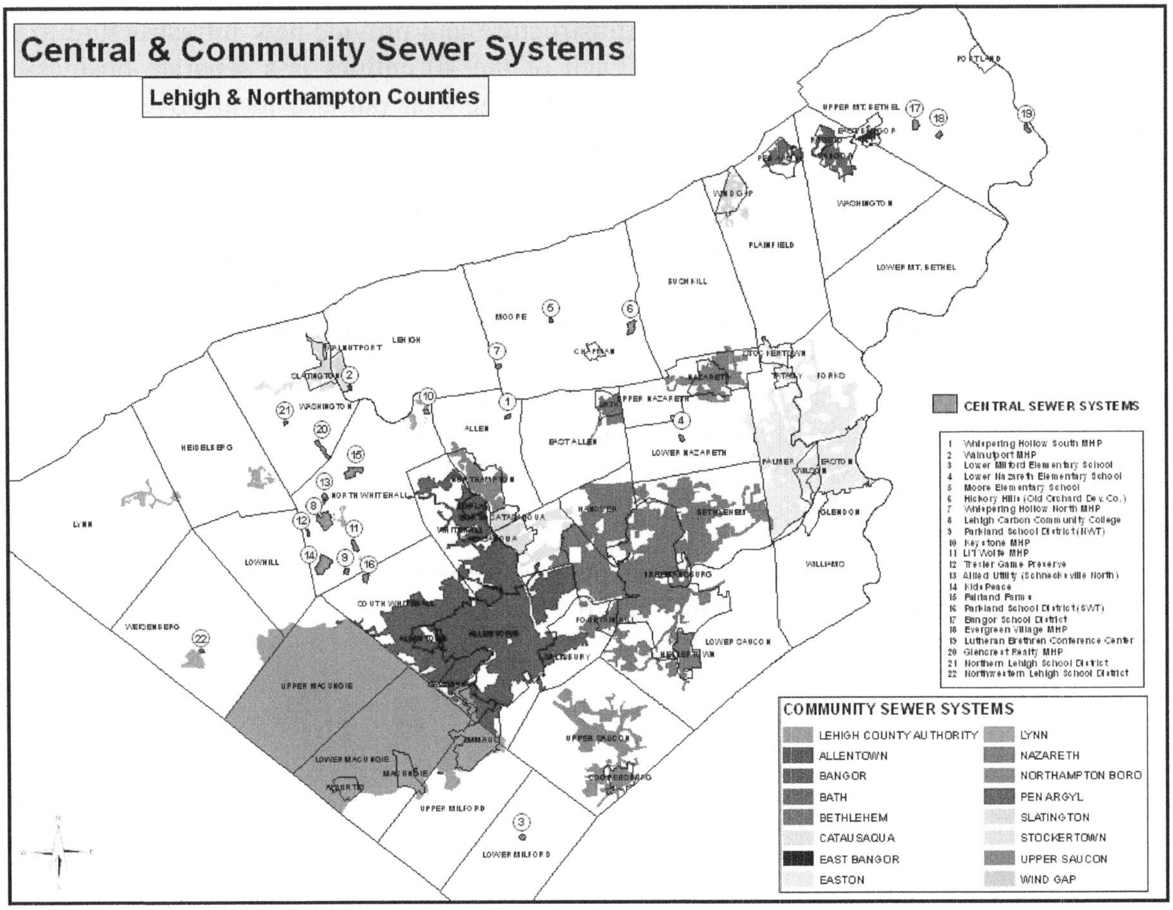

Figure 2.2 Lehigh Valley central and community sewer systems

It is natural to think of water and wastewater service as being a very physical enterprise, involving facilities that are distributed over large geographic areas and physical processes related to hydrology and water quality. However, it is also an inherently financial enterprise due to the large amount of capital investment required to meet the physical challenges presented by obtaining, treating, and transporting water. It is, in fact, the most capital intensive of all utility services—requiring more than twice as much capital per dollar of revenue than electric power, even with its billion dollar power plants.

In earlier times, when water sources were abundant, regulatory requirements were modest, and the infrastructure was new, the inherent operating and financial challenges of the business were not as apparent and it was common to see a proliferation of utility enterprises created as a part of local governments. In the period ahead, many of the cost drivers are reversed. These utilities must now face resource constraints, more demanding regulatory requirements, and (for some) the first-ever need to re-invest significant capital in the replacement of expensive infrastructure. Such demanding changes in business environment suggest there is potential benefit to be gained from a regional organization such as that in Charlotte-Mecklenburg. At a regional scale, there are significant cost advantages in meeting the inherent physical challenges of the business as well as significant risk management advantages in meeting the inherent capital financing challenges of the business. With these changing cost drivers, the 40 independent entities in the Lehigh Valley that have been adequate to meet the task up until now will be increasingly disadvantaged by their lack of access to these benefits of regional collaboration.

FINANCIAL RISKS OF WATER AND WASTEWATER CAPACITY MANAGEMENT

Suburban and rural jurisdictions are confronted by growth in demand for water and wastewater service. They also must make risky choices regarding how much capital to invest in water and wastewater treatment plant capacity. Many jurisdictions are nearing a decision about upgrading treatment facilities that were constructed during the 1970s, 1980s, and early 1990s. The cost of replacement can greatly exceed that of the original investment. Determining how much capacity to buy in order to allow for new growth is a risky decision. While the growth in the Lehigh Valley is a strong trend that seems likely to continue, it is never possible to tell for certain how growth will occur. Recent U.S. Census data for the Lehigh Valley shown in Figure 2.3 indicate that growth has materialized more slowly in some areas where it seemed assured and yet has occurred more swiftly in others where it was not expected to be so vigorous.

In a number of instances across the Lehigh Valley, there are existing collaborative arrangements between jurisdictions to share in the cost of treatment capacity at small scale facilities. In general, these agreements have a mixed track record. Some have led to disputes that arise when one party or another is affected by changes in growth and demand that leave it with either too much or too little capacity. It is hard enough to estimate capacity needs in a single jurisdiction facing these growth issues. With two or three jurisdictions involved, the odds are greater that at least one of them will guess wrong. At the level of an individual township, the difference between being right and being wrong in assessing capacity needs can swing on the fate of a few development projects. Within a small treatment facility, this can amount to a sizable percentage of the total capacity cost.

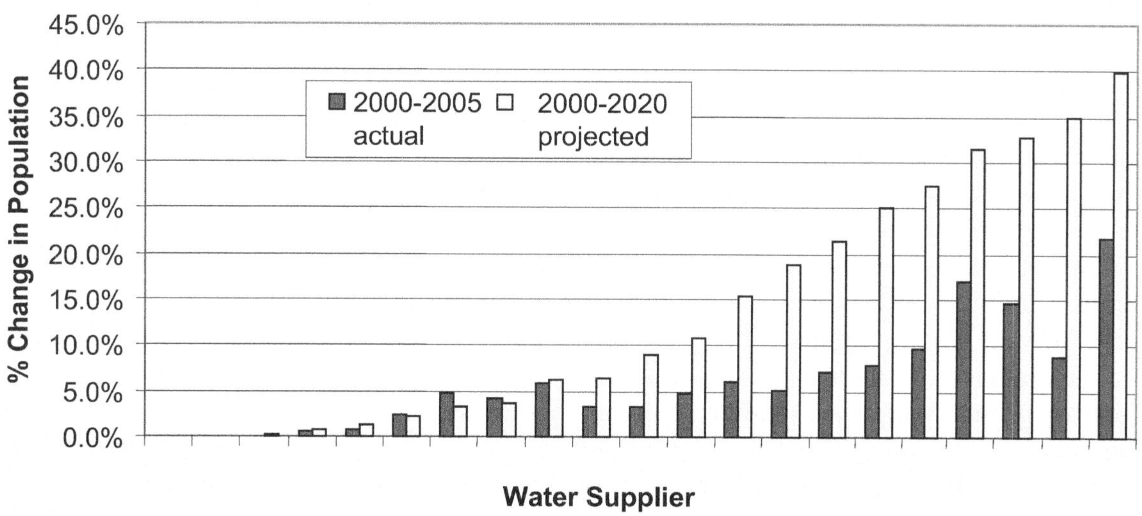

Figure 2.3 Lehigh Valley growth forecasts compared to actual growth

When smaller communities make treatment capacity decisions, they are making firm financial commitments that, in turn, give them a clear set of incentives to make their bets on capacity needs turn out to be the right ones. Throughout the Lehigh Valley, many municipal owners of smaller treatment plants are obtaining a significant share of their revenues from new connection charges that are used to recover the capital investment. These cash flows can create a financial dependence on growth and an incentive to sustain it. Once the owner and the wholesale customers (if any) reach build-out, these cash flows will dry up and the true costs of treatment plant ownership will be recognized.

Even if it were simpler to define just the right amount of treatment capacity, there are additional types of financial risks that must be borne by the owners of treatment capacity and their wholesale customers. Unexpected events are remarkably common among small utilities in the Lehigh Valley. Several have experienced incidents of water source contamination in which they have had to scramble to cover millions of dollars in unexpected response costs. Even when some portion of such costs is recoverable from third parties, it cannot be recovered without extensive outlays upfront, legal expenses, and risk. Many have had the experience of either gaining or losing a large commercial or industrial customer which can upset carefully planned capacity allocations and projected revenues. Many have had their expected treatment capacity modified as a result of changes in the U.S. Environmental Protection Agency (USEPA) regulations or as a result of unexpected changes in other variables affecting their permitted capacity. All of these influences can upset the most careful financial plans and there is little room for error in a small scale utility business.

As an alternative institutional approach, the authors suggest consideration of one or several larger institutions that could be formed by participating jurisdictions under the Pennsylvania Municipality Authorities Act to serve the specialized purpose of owning and operating treatment plants and related aspects of water infrastructure (e.g., supply sources) and wastewater infrastructure (e.g., interceptor sewers). The objective would be to relieve individual jurisdictions of the financial risks of owning and operating treatment capacity and essentially pool those risks across a larger region. While the impact of a half million gallon per day change in capacity arrangements can have a significant financial impact on the owner of a 5 million gallon per day plant, it would amount to almost nothing if spread across a larger region.

The authors suggest consideration of either a single authority covering the entire Lehigh Valley or three sub-regional entities centered about Allentown, Bethlehem, and Easton. The recent restructuring of arrangements for water service in Easton and its nearby suburbs under a single authority is an example of the type of larger entity envisioned in the sub-regional alternative. Gathering the treatment plant and related functions of water and wastewater service under a larger authority covering a wider region would not only relieve smaller jurisdictions of the financial risks of owning and operating treatment capacity, it would also provide a healthful separation between the business of running a utility and the financial operations of a municipality. It would eliminate any potential linkage between connection charge revenues and municipal finances, reducing the risk that water or wastewater financial concerns would affect municipal growth policies. Growth policies could remain a local matter through various mechanisms. For example, the regional authority could be required to adhere to capacity expansion and line extension policies specified by the participating municipalities that each maintain sovereign control over land use within their borders.

As stand-alone utilities, the older Cities of Allentown, Bethlehem, and Easton face financial risks of making significant upgrade and capacity decisions in existing large-scale treatment plants. Those plants serve not only their own residents, but also a number of adjacent suburban communities as wholesale customers. The financial risks to the older cities are magnified by demographic shifts that are causing them to lose population and hence the tax base and customer base to support such large investments. Their arrangements with suburban jurisdictions are anchored in old agreements that worked well in the era when the manufacturing sector was the main economic driver in the Lehigh Valley and city finances were more robust. But these agreements have been under strain more recently. Today, the city books are already loaded with water and wastewater debt, pre-empting other municipal investments (Figure 2.4). Both Allentown and Bethlehem have had their bond ratings lowered recently. The prospect of taking on still more water and wastewater debt and more of the risks of treatment plant operation, in part for the benefit of other jurisdictions, is not as attractive as it once may have been. Suburban wholesale customers are meanwhile increasingly anxious about the ability of the older cities to provide the capacity needed for the future on a reliable basis. In some instances, they have begun to evaluate options for developing their own capacity, leaving the city system more financially strapped from the potential loss of revenues. Given these adverse trends, the current arrangements for running these water and wastewater businesses may not be sustainable and can only add to the financial stress on the older cities.

It is in response to this dilemma that the recent restructuring of water service in Easton was brought about. By turning the water business over to an authority, the city has been freed of the operational risks and its books are free of both past and future water debt obligations. This will free city capital resources for other projects.

This seems like the right direction for Allentown and Bethlehem also. Throughout the 1990s, both cities were compelled to make large investments in drinking water treatment capacity due to new regulations. On the wastewater side, both cities have been plagued by the effects of infiltration and inflow into sewers during rainfall events that reduce the available treatment capacity—a problem that will require still more capital investment. In each case, a new

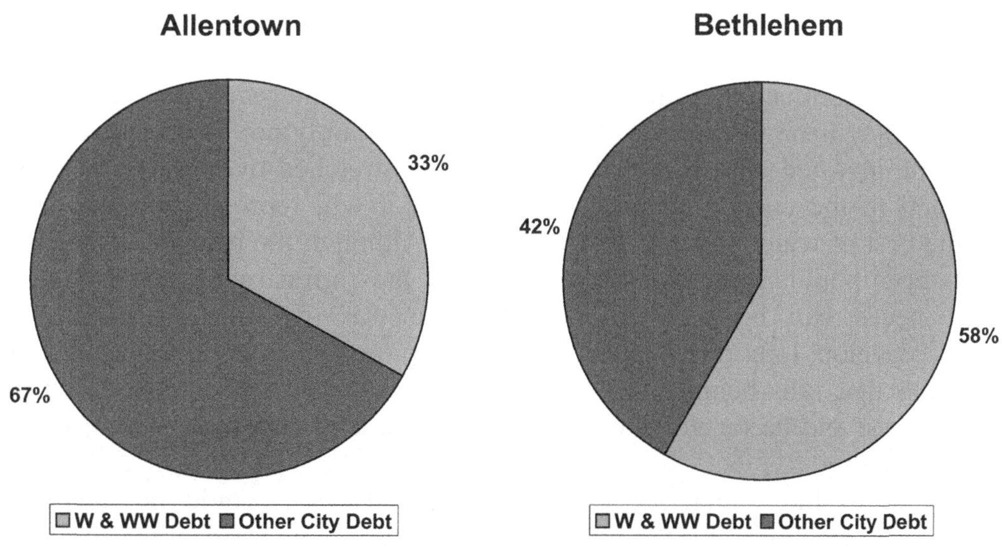

Figure 2.4 Water and wastewater debt as a share of total city debt

authority could be formed that would include representation on the governing board from the city as well as suburban communities. The past and future water and wastewater debt could be removed from the city books and transferred to such an authority along with the operating risks of running the facilities.

Suburban communities served by Allentown and Bethlehem would also benefit from conversion to authority status. Under an authority structure, they would have input into the chartering of the organization as well as representation on the governing board. This would enable them to oversee the management of the business and obtain greater assurance of getting the capacity they require. In fact, the entire region could benefit from removing the burden of water and wastewater debt from the city books in Allentown and Bethlehem. By liberating debt capacity for reinvestment in these older cities, an urban development alternative could be created that might provide a counterweight to suburban growth, restoring balance to the real estate market and easing the rush to the outer suburbs.

The idea of creating sub-regional authorities to own and operate treatment plants and related facilities could be scaled up to the next level—creating a single authority that would be responsible for all treatment plants, large and small, throughout the Lehigh Valley. This would provide the greatest opportunity to pool the operating risks of owning treatment capacity across the region as well as producing additional operating efficiencies.

FINANCIAL CHALLENGES OF REPLACING AGING PIPES

All across the country, communities are awakening to the realization that they are at the "dawn of the replacement era" in water and wastewater utility service (Cromwell, Speranza, and Reynolds 2001). Pipe networks that have had exceptionally long service lives are finally coming due for replacement, as evidenced by an increased level of concern for main breaks. As a result, utilities are faced with the need to begin ramping up significant new capital expenditures to rehabilitate or replace older pipes. A prudent utility will begin now and continue increasing pipe reinvestments for the next several decades. This will add significant new costs to utility bills that have never been experienced by previous generations of utility customers because the pipe infrastructure has lasted such a long time. Most utilities in the Lehigh Valley have spent only minimal amounts on replacement investment to date.

The authors estimate that the total cost of replacing water and wastewater pipe networks in the Lehigh Valley is on the order of $4 billion. It is certainly not the case that such enormous sums need to be invested all at once; the cost will be stretched over many decades, extending through the end of the current century. Nonetheless, it will represent a large and permanent increase in the cost of water and wastewater service. Although the problem posed by aging pipe networks has been popularly characterized as part of the "infrastructure crisis" in America, the reinvestment needs can be met in a manageable way if utilities carefully study their infrastructure, develop long-range plans, begin ramping up expenditures now, and stay the course. A utility that falls behind will forfeit the opportunity to phase these cost increases in gradually over time and be faced instead with more dramatic cost increases needed to catch up later.

It may seem like replacing aging pipe infrastructure would be mostly a financial challenge for the older cities, but the need to ramp up replacement spending is equally pressing in the suburbs. This results from the odd fact that the oldest pipes are not necessarily the ones that need to be replaced first. The oldest "pit cast" iron water pipes that were placed in the ground in Easton, Bethlehem, and Allentown 100 years ago were considerably over-designed and many have substantial life left in them. In contrast, the next generation of "spun cast" iron pipe that was installed from the 1920s through the 1960s has proved to be much less durable and is the source of most of the main breaks experienced today. As shown in Figure 2.5, the suburbs were growing vigorously during this time and, as a result, there is about as much of this pipe in the suburban jurisdictions as there is in the older cities—if not more. Almost all of the utilities interviewed in this study cited spun cast pipe as their greatest infrastructure concern. On the wastewater side, the most pressing pipe infrastructure needs concern concrete interceptor sewers that were extended from the older cities to the suburbs in the 1960s and 1970s. Depending on contractual arrangements, replacing or rehabilitating these interceptors may be the combined responsibility of city and suburban users of the major wastewater treatment plants.

A body of best practices referred to as "asset management" has been developed to describe the systematic application of research and planning techniques to optimize the rate of infrastructure replacement. A well-planned replacement program presents significantly less financial risk and customer impact than a poorly planned one. The intent is to avoid a situation where the utility is incurring increased operating expenses to fix an increasing number of breaks at the same time it is trying to raise more capital to replace the same problem pipes. Asset management practices are best applied at a large scale, however, where the scale of the assets involved makes the cost of the research and planning efforts involved affordable by comparison to how much they contribute to a more cost-effective solution. This is a very fruitful area for enhanced regional collaboration in the Lehigh Valley.

A region-wide collaboration in asset management could conduct research and planning for the needed investments in pipe replacement. This could be approached in several different modes of collaboration. A few regions around the country have formed consortiums of neighboring utilities with similar vintages of pipe materials to achieve the purpose of obtaining an economy of scale in asset management. A loose consortium is limited by the strength of the commitment of its members. To assure results from such a collaboration requires more of an institutional structure.

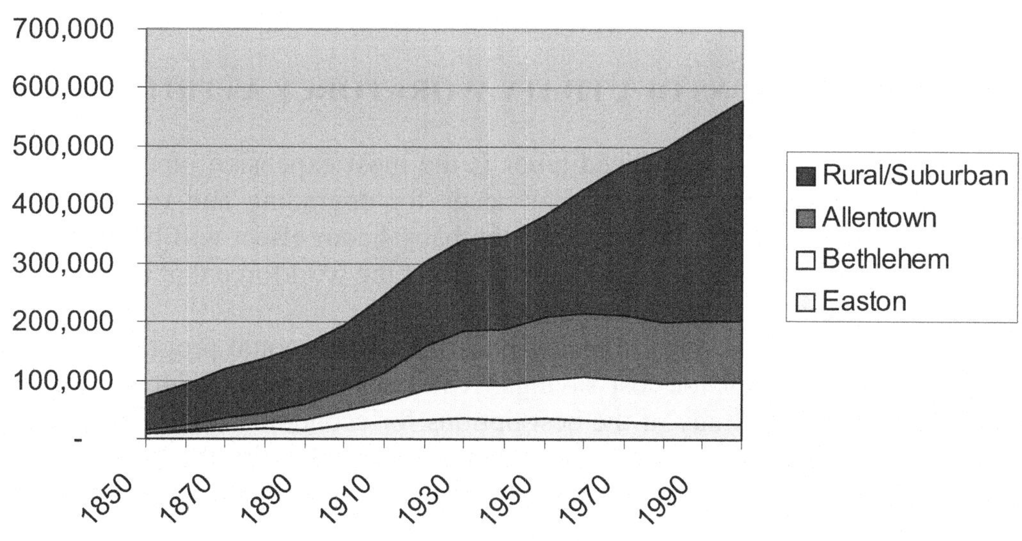

Figure 2.5 Historical population growth in the Lehigh Valley

Once again an authority can be the mechanism for utilities in the Lehigh Valley to combine the asset management planning responsibility for all the water and wastewater pipes. This would enable the Lehigh Valley to obtain the economies of scale that are inherent in asset management planning [e.g., use of Geographic Information Systems (GIS) tools] and have greater assurance that the effort will produce results. In addition, there are many closely allied tasks that could be very effectively carried out by such an authority. Foremost among these would be financing and procurement. Since planning and financing are always tightly related functions, the authority could sell revenue bonds to raise the required capital to be recovered from member utilities through a surcharge. This would have the additional benefit of avoiding a large demand on debt capacity at the local level. Over the longer-term, this could also provide a means of gradually transferring ownership of pipe infrastructure assets to the larger regional entity. And, of course, another way to achieve the same ends would be to pursue either a regional or subregional consolidation of utilities that would encompass pipe infrastructure as well as treatment facilities.

A regional authority could assume planning and financing functions as well as actual operation and maintenance of pipe networks in member jurisdictions. This might be accomplished through the transfer of key local staff to the authority in order to capitalize on their deep knowledge of the existing pipe networks. Additional economic benefits would result from being able to gradually convert the pipe systems in the region to a common set of standards and specifications. Individual jurisdictions would benefit from having the support of a large organization comprised of many experienced operations staff rather than being wholly dependent on a few people. The consolidation would also provide a strategy for capturing the invaluable knowledge of key operations staff in the planning process.

It does not take too much extrapolation to see how such a region-wide authority could develop a remarkable concentration of knowledge about the pipe assets from which to truly optimize the cost of maintaining them in good condition. Working against a $4 billion price tag, the payback potential is significant. The crowning touch would be to combine GIS mapping technology used in the planning process together with a region-wide customer call center that could improve responsiveness in the event of pipe failures or other performance problems. A centralized point of contact for utility customers also would capture this performance data as another input to the research and planning processes of asset management.

FINANCIAL IMPLICATIONS OF UTILITY WORKFORCE ATTRITION

After capital facilities, specialized labor is the most expensive input in the provision of water and wastewater service. Just as major needs for upgrading and replacement of capital facilities come due over the next dozen years, the baby boom effect will be causing utilities all across the Lehigh Valley to lose a crucial component of the workforce that is integral to efficient planning and management of such capital programs.

It is well established that the planning phase of major capital programs is the phase that offers the greatest opportunity for cost savings. It is much more difficult to obtain economies once designs have been set. Many of the best options for economizing on future or replacement facilities will be most evident to those who have been working with the local infrastructure for many years. Their accumulated knowledge is invaluable and scheduled for retirement soon.

Water and wastewater utilities require a highly skilled workforce that is becoming more and more difficult to recruit. The regulatory performance requirements imposed on water and wastewater systems have become enormously more complicated since the present generation of managers and operators began working in the field. Across the industry, there is a serious concern over attracting sufficient engineering talent into the field on public sector wage scales (that often are not competitive with private sector offerings) to a shrinking pool of graduating engineering students. Many small municipalities already outsource the engineering function to consulting engineers rather than maintain staff capability in the utility. But consulting engineering firms are having the same difficulties in recruitment with the same effect on labor costs. The same problem exists in attracting and maintaining highly skilled operations staff on public sector wages when the private sector is expressing a growing demand for such skilled labor. Throughout Pennsylvania, data from the Bureau of Labor Statistics confirm that wages for water and wastewater operators are accelerating faster than inflation.

In confronting the workforce challenge, one of the major disadvantages faced by small municipal utilities is that their existing staffing strategy often reflects a blending of utility needs and other municipal needs. So, a borough or township may have two or three people, or a dozen people, who work in the utility, but who also fulfill other municipal functions. This strategy has the benefit of covering part of the cost of a number of positions from water and wastewater revenues rather than from general funds. But this staffing pattern is a historical artifact that reflects labor market conditions of the past and will be increasingly difficult to replicate in the more competitive labor market that is now upon us. Municipalities will find it increasingly difficult to attract and maintain the same number of people with the skill level required for utility operations in today's complex regulatory environment. This will also make it tougher and tougher to allow such valuable labor to be diverted to other municipal functions.

A larger entity such as a sub-regional or regional water and wastewater authority would have much more leverage in meeting this workforce challenge. Not only could it undertake recruitment and training with the economies of a larger-scale organization, but it could make much better use of both the existing and future labor forces. By isolating these utility functions in a stand-alone authority, removed from diversions into other functions of municipal government, it could be possible to deploy valuable labor resources much more efficiently and leverage that deployment with information and communications technology to provide water and wastewater service with a smaller overall workforce.

One of the obstacles to creating larger entities arises from the implied intention to reduce the labor force. In this unique window of time over the next dozen years, the pending wave of retirements among the baby boom generation will make it possible to transition painlessly to a larger entity without the need for layoffs, since natural attrition will be more than adequate. Indeed, it is likely that many members of this retiring generation will need to be retained in the workforce as long as possible in order to harvest their knowledge and tap into their leadership skills in forming a new future for the Lehigh Valley.

CHAPTER 3
DATA AND METHODOLOGY

OVERVIEW

The viability and sustainability of a business is represented in a business plan that includes a financial forecast to establish expectations for the performance of the enterprise. The same methods are applied here in comparing alternative futures for publicly owned water and wastewater utility businesses in the Lehigh Valley.

As discussed in chapters 1 and 2, such long-term forecasting is not typical in publicly owned utility businesses, creating a blind spot regarding changes in the business environment. As described below, forecasting models for Lehigh Valley utilities were created by first developing an understanding of the performance of these utility businesses under the status quo institutional arrangements (using data from year-end 2005) and then projecting the impacts of changing cost trends and business risks on that status quo baseline. These "status quo scenarios" establish expectations for the existing utility businesses in 2020 in terms of the household cost of water and wastewater service.

For comparison, the forecasting models are adapted to reflect regional consolidation scenarios as alternatives to maintaining the institutional configurations of the status quo. The results indicate a different set of expectations in terms of the household cost of water and wastewater service in 2020.

In all scenarios, it is assumed that the utility businesses will meet the challenges posed by changing costs and risks over the forecasting period, but the cost-effectiveness with which they can be met differs based on the type and size of the business. The result of this differential is reflected in the 2020 household cost which is therefore the basis for comparison between the scenarios.

DATA COLLECTION

Invitations to meet with the project team were extended to all 40 utilities in the Lehigh Valley. About half accepted the invitation and participated in one-hour meetings designed to obtain and review recent operating and financial reports of the utilities and discuss the experience and prospects of the owners and operators in running the systems. The interview process provided some deep insights and revealed a number of repeating patterns among the utilities.

Financial reports indicated that the utilities in the Lehigh Valley are generally solvent on a year-to-year basis. Utilities in the older Cities of Allentown, Bethlehem, and Easton (prior to absorption of the water system into Easton Suburban) have large debt loads as a result of supporting large regional treatment facilities. Utilities in the outlying growth areas tend to have large cash flows from new development that are currently providing a strong revenue stream. However, due to transfers between utility and local government accounts, it was not always possible to determine from the financial statements the extent to which these cash flows were being devoted to utility services versus general government services.

It also appears that many utilities are regarding cash flows from new connection fees as operating revenues available to cover operating expenses rather than as capital recovery. This will keep rates lower in the short-term, but could lead to significant rate increases once the service area approaches build out.

The major impression that was reinforced many times during the interview process was the high degree to which individual utilities are taking on substantial financial risks by being the owners and operators of treatment capacity. Treatment capacity is inherently very expensive to own and operate in the first place; and then, a lot of things can go wrong to make it more expensive. And, almost all of those things have been experienced by utilities in the Lehigh Valley over and over again in the recent past.

Once a level of treatment capacity has been selected, the economics of the utility will hinge on its ability to maintain enough flow to utilize that capacity acceptably. The authors encountered a number of small utilities with idle capacity for sale—leaving current rate payers to cover the bond payments on their own while waiting for the growth to materialize. But there were also as many cases where capacity needs have been underestimated relative to the growth and neighboring jurisdictions have ended up in court to settle disputes over the capacity that exists, resulting in significant legal costs on both sides of the argument.

In addition to the unruly nature of growth patterns, numerous other unlucky surprises can come along to upset the best laid plans for building and sharing treatment capacity. Several large and small utilities have had to overhaul surface water treatment facilities to add filtration in compliance with Safe Drinking Water Act standards, adding enormously to the cost of treating surface water. Many well-supplied systems in the Lehigh Valley continue to be subject to the risk of being redefined as "surface influenced supplies" by state regulators. Many small wastewater treatment facilities within the Lehigh Valley are fearful that state regulators will designate their receiving waters to be of trout stream caliber and therefore require more expensive advanced treatment under the state's Clean Streams program.

Infiltration into wastewater interceptors and other large collection mains has also caused a number of utilities (notably, Allentown and Bethlehem) to reduce the amount of treatment capacity available to conform to the amount that the state will approve for their facilities. This requires either expensive rehabilitation and replacement of large pipe infrastructure or expensive over-design of treatment capacity.

Other types of surprises were also reported. Several small water utilities have experienced severe well contamination incidents caused by releases of hazardous materials into groundwaters. These episodes have cost small utilities millions of dollars in legal fees, emergency response costs, and remediation expenses. Several utilities also reported the sudden appearance or disappearance of a large commercial/industrial customer. Whether coming or going, the impact on operations is often sudden and of significant magnitude.

STATUS QUO MODELING

Using the data collected from participating water and wastewater suppliers, simplified revenue requirements models were developed for 10 generic types of utilities:

- Small water
- Small water distribution only
- Medium water
- Large groundwater
- Large surface water

- Small wastewater treatment
- Small wastewater collection only
- Medium wastewater treatment
- Medium wastewater collection only
- Large wastewater treatment

The purpose of the models was to estimate the revenues that each type of utility would need to collect in order to be financially viable. Financial viability was defined as being able to pay all operating expenses, meet all debt service requirements, and have remaining revenues consistent with current industry standards. Industry standards were defined as having a ratio of revenues to expenses of at least 1.30 and having debt service coverage of at least 1.50 times.

A ratio of revenues to expenses (known as the operating ratio) of at least 1.30 times has been shown to be correlated with a water system's ability to reliably meet all regulatory requirements and provide the public with a safe and reliable water supply (Cromwell et al. 1997). This is confirmed by data compiled by Fitch Ratings for more than 150 public water and wastewater utilities showing that the median operating margin for an investment-grade water or wastewater utility is 34% (Fitch 2007). The operating margin is revenues minus expenses divided by expenses, which is equal to the operating ratio minus 1. Thus, an operating margin of 34% is equal to an operating ratio of 1.34.

A debt service coverage ratio is a measure of the cash available to the utility in excess of its debt service payments. It is measured by comparing the utility's net cash revenues [that is revenues minus operating expenses plus non-cash expenses (which are usually made up of depreciation expense and miscellaneous amortizations)] to the utility's debt service payments. A debt service coverage of 1.50 times is consistent with the median performance of the lowest-rated investment-grade water and wastewater utilities. Specifically, the Fitch study shows that the median coverage ratio for all investment-grade water or wastewater utilities in that study was 1.8 times, while the ratio for A-rated utilities (the lowest rating in the Fitch study) was only 1.5 times (Fitch 2007).

Debt service is calculated by assuming that the net value of utility plant (gross plant less depreciation reserve, grants, and contributions) is financed entirely with debt. While this may not be a valid assumption for all publicly owned utilities, it essentially recognizes the opportunity cost (or cost of capital) associated with internally generated financing. In effect, then, it assumes that internally generated funds should earn a cost of capital equivalent to the cost of debt. This is a very conservative assumption, as most rate-setting methodologies recognize that internally generated funds (known as equity capital) should earn a return that is higher than the cost of debt (AWWA 2000).

Operating expenses were calculated as the sum of the following:

- Salaries and wages
- General and administrative
- Utilities
- Materials and supplies
- Miscellaneous services
- Purchased water
- Depreciation and amortization

Each of these costs was projected separately, as described in appendix A.

The net value of utility plant was calculated as the sum of the following:

- Net capital assets as of 2005
- Pipe replacement investment
- Other plant replacement
- Plant/well upgrade, addition #1
- Plant/well upgrade, addition #2 (in higher-growth scenarios)
- Plant/well upgrade, risk (in risk scenarios)

Each of these plant investments was assumed to occur over time between 2005 and 2020, so the remaining value in 2020 reflected the depreciation of assets between their installation and year-end 2020. The method for projecting these costs is described in appendix A.

Both plant costs and operating expenses were varied with the amount of population growth a system was projected to experience. Population forecasts at the municipality level were prepared by the Lehigh Valley Planning Commission in 2007 (LVPC 2007). These forecasts were used to develop "growth" and "no growth" scenarios for each type of utility. For example, small water utilities in "no growth" areas are projected to have their populations grow, on average, by a total of 1.4% between 2005 and 2020. In contrast, small water utilities in "growth" areas are projected to have their populations increase by an average of 24.2% over the same time period. From the population forecasts, it was determined that there would be five small water utilities in "no growth" areas and six in "growth" areas. The specific methodology for varying costs with population growth is discussed in appendix A.

Debt service was calculated by assuming a 20-year bond with semi-annual interest payments and an interest rate based on a projection of AA-rated utility bond costs in the year of the first plant upgrade. The projected interest rate on utility bonds was prepared by the U.S. Energy Information Administration as one of the macroeconomic indicators used in its long-term energy price forecasting model (EIA 2006). This interest rate was then reduced by 35% to account for the tax-exempt nature of public financing that would be used by the Lehigh Valley water and wastewater systems. This is a conservative assumption since most of the smaller utilities in the Valley have bonds with a lower rating than AA, and therefore would have a slightly higher interest cost.

The simplified model calculates required revenues as the greater of the following two equations:

$$\text{Operating expenses} \times 1.30 \tag{3.1}$$

$$(\text{Debt service} \times 1.50) + \text{operating expenses} - \text{depreciation expense} \tag{3.2}$$

To facilitate the comparison of systems of different sizes, required revenues were then divided by the amount of water sold to customers, to develop a cost per thousand gallons. To help the general public understand the costs, this amount was multiplied by 60 to represent the typical water consumption for a residential customer (or household) in the Lehigh Valley (60,000 gallons per year). This typical household cost was then used to compare the impact of various scenarios on the cost of water and wastewater service in 2020.

MODELING RISK

One important aspect of projecting future costs is often overlooked: the element of risk. Risk can be manifested in many ways. For a water system, risk can occur as the contamination of a well, a major dispute with a purchased water supplier, the loss of a major industrial customer, a flood, or a similar event that could have a major impact on the utility's revenues, expenses, or capital requirements. During the project team's interviews with Lehigh Valley utilities, each of these types of major events has affected utilities in the Lehigh Valley during the past few years.

For purposes of the simplified model developed for this project, risk was modeled as an unexpected major plant addition. In effect, it has the same cost impact as a new well being constructed.

Certainly, this is a simplifying assumption. As noted, risk could be manifested as an abrupt change in revenues or expenses, not solely as a capital investment. For the type of simplified model developed for the Lehigh Valley, however, this appears to be an appropriate way to recognize the potential impact of unforeseen events on a utility's costs.

It was assumed that 50% of the utilities would experience a "risk event" during the 15-year forecast period. In effect, this assumes that such an event would occur once every 30 years for each utility. Given the prevalence of the events that have occurred in the Lehigh Valley over the past few years (including two major floods, at least three major (and unrelated) well contamination events, at least two major wholesale purchase disputes, and several industrial plant closings), it appears to be conservative to assume that such an event would affect each utility only once each 30 years.

The risk assumption was modeled by calculating the cost impact in 2020 of having a new well constructed half-way through the 15-year forecast period (2013) for each type of water utility. A plant addition of similar magnitude is assumed for wastewater utilities and surface water treatment water utilities. For purchased water systems, it was assumed that their supplier experienced the unforeseen event.

Because the plant addition depreciates each year, assuming that all risk-related investments occur at the midpoint of the projection period has roughly the same aggregate impact as assuming that the events occurred ratably over the projection period. Thus, while it is certainly a simplifying assumption, it is a reasonable assumption for this type of modeling because the overall cost impact in the year of interest (2020) will be approximately the same as assuming that the "risk events" are spread out evenly over the forecast period.

The model then assumes that 50% of the utilities of each type were "unlucky" (that is, they experienced the "risk event"), and that 50% were "lucky" during the 15-year period. Thus, the risk-enabled analysis further subdivides the categories of water systems (e.g., "lucky" small water, "unlucky" small water, "lucky" medium water).

REGIONAL SCENARIO MODELING

Two regionalization scenarios were modeled. The first is a sub-regional approach that would create three large water and wastewater supply entities in the Lehigh Valley. The three sub-regions roughly correspond to areas surrounding the three major cities in the Lehigh Valley: Allentown, Bethlehem, and Easton. Outlying areas in the northern portion of the Lehigh Valley would remain independent under this scenario. The second scenario is a complete regionalization scenario where all water and wastewater suppliers would be owned and managed by a common regional entity.

Importantly, neither scenario includes the physical consolidation or interconnection of any water systems. Any savings from regionalization are assumed to come from increased efficiency and risk sharing. Certainly, future engineering studies may show that physical consolidation and interconnection might achieve additional savings. But given the nature of this project, no such savings were projected. This is a further conservatism in the analysis.

Cost forecasts under each regional scenario were developed using the same revenue requirements model. Two significant types of efficiencies were modeled from regionalization: (1) reductions in operating expenses, and (2) efficiencies in infrastructure replacement planning and implementation.

Moving to sub-regional organizations was estimated to reduce operating and maintenance expenses by 20% in 2020, as compared to the status quo entities. Savings of this magnitude are consistent with a strategy of leveraging the workforce with advanced automation and instrumentation. While the introduction of these workforce changes implies reductions in the workforce, the next 15 years represents a unique period during which a natural wave of retirements will take place within the World War II baby boom generation. This natural wave of attrition represents a one-time opportunity to make a painless transition to modern management practices while the baby boomers are still around to lend their knowledge and experience to the process. In addition, nearly every water and wastewater supplier in the Lehigh Valley is facing a challenge with its aging management workforce. By 2020, nearly every general manager or public works director in the region will reach retirement age, as will many of the managers, engineers, lawyers, and accountants whose functions could be consolidated to serve several utilities. Consequently, the 15-year period studied also presents a unique opportunity to obtain efficiencies at the management (or "back office") level without adversely affecting anyone's livelihood.

Transitioning from sub-regional organizations to a single regional entity was modeled to reduce operating and maintenance expenses by an additional 10%, or by 28% as compared to stand-alone operations.

Savings of as much as 30% are consistent with a recent study by Shih and colleagues, who found that economies of scale on the order of 10% to 30% were achieved when the size of a water system doubled. They also found that "the greatest economies of scale exist in the cost categories of capital, outside services, other, and materials. … This suggests that larger systems may be relatively better than smaller systems at bargaining for and receiving outside services and materials at lower costs. In addition, it is important to note that these gains do not necessarily depend on water systems becoming physically interconnected" (Shih et al. 2006).

Decreased costs associated with infrastructure replacement were also modeled. Data compiled by the American Water Works Association (Cromwell, Speranza, and Reynolds 2001), indicate that the typical water utility replaces approximately 0.7% of its water main infrastructure on an annual basis. A popular rule of thumb in the industry encourages utilities to strive for a 1.0% annual replacement rate. This has been shown to be excessive, however. Tremendous efficiencies are achievable through modern practices of asset management (Cromwell et al. 2006) that enable utilities to extend the useful life of pipe assets and manage them more economically while still sustaining desired service levels. For modeling purposes, it was assumed that a stand-alone utility would replace 0.7% of its pipe infrastructure each year, a sub-regional entity would be able to reduce the replacement rate of 0.35%, and a regional entity would reduce it further to 0.3% per year.

Finally, and importantly, the consolidation scenarios also assume the spreading of the impact of "risk events" among a larger group of customers. Thus, while the same number of "unlucky" events are assumed to occur across the region over the period, the cost impact of those events on individual customers is reduced because the cost is shared among all customers in the regional or sub-regional entity. In effect, the larger entities serve as a risk (or insurance) pool to spread the risks of being "unlucky" among a much larger group of customers.

CHAPTER 4
RESULTS

WATER

Status Quo Projected to 2020

In 2005, the typical residential water customer in the Lehigh Valley paid a water bill of between $140 and $290 per year (based on consumption of 60,000 gallons per year). In total, Lehigh Valley water suppliers serve approximately 165,000 residential customer accounts. At present, it is estimated that the total annual water bill for those residential customers is on the order of $37 million per year.

Using the water cost model developed for this project, and assuming no structural changes in the organization and operations of the Lehigh Valley's water suppliers, it is estimated that these costs will increase dramatically by 2020. Most customers are likely to see their water bills double or triple by 2020, while some customers (those whose utilities experience source contamination, a natural disaster, or other "risk event") could see their bills increase four-fold or more compared to 2005.

It is estimated that the average water cost will increase from its current Lehigh Valley-wide average of about $220 per year to approximately $510 per year in 2020. This would result in a total residential water cost of approximately $91 million in 2020, an increase of nearly 150% over existing costs, as shown in Figure 4.1.[1]

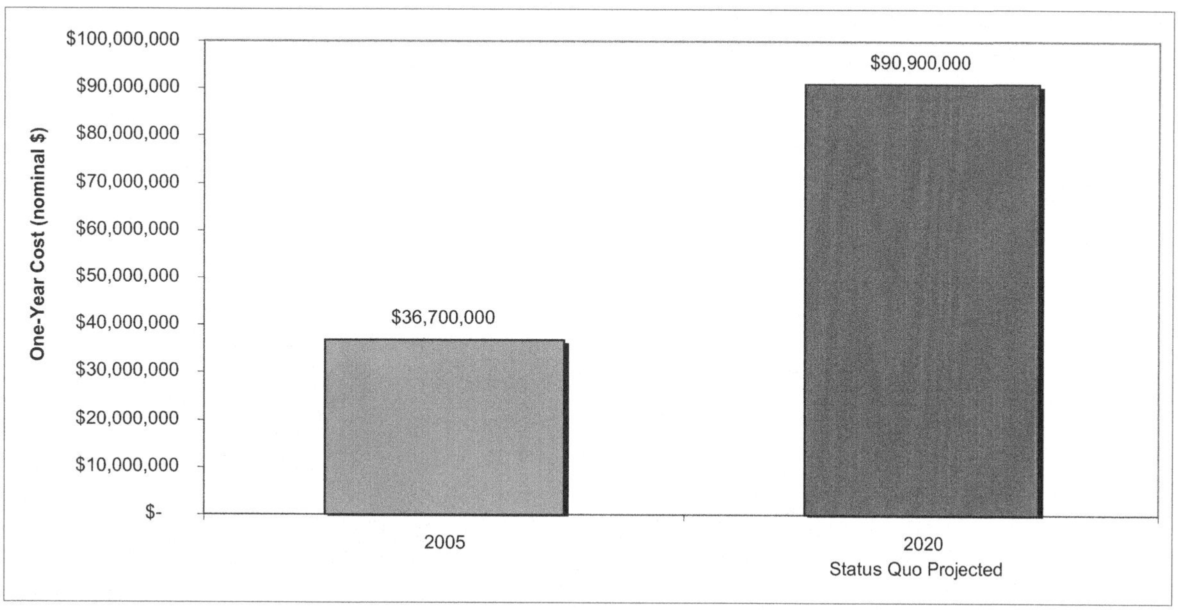

Figure 4.1 Total residential water costs in 2005 and 2020 status quo projected

1. All cost projections in this chapter are presented in nominal dollars, including the projected impact of inflation.

While numerous cost drivers were modeled, the projected cost increase is largely the result of three factors: (1) the inefficiency of capital replacement and rehabilitation in smaller utilities; (2) the inability of smaller utilities to spread the risk of contamination, facilities failure, or other significant adverse events; and (3) general inflation.

Further, the range (or dispersion) of annual household water costs will change dramatically. Today, customers of different water suppliers pay costs that are relatively close to each other throughout the Lehigh Valley—ranging from about 30% above the average to about 35% below the average. By 2020, however, costs will diverge substantially, so that some customers will see water bills that are 75% higher than the average in the Lehigh Valley, while others might pay costs that are as little as 45% below the Lehigh Valley-wide average. Thus, it is projected that the range of water costs by 2020, under the status quo scenario, would be from a low of $280 per year to a high of $890 per year.

The most extreme results will be experienced by customers of smaller water utilities. Indeed, the model results show that the smaller the utility, the higher the costs are likely to be in 2020. Figure 4.2 illustrates these results using a bubble chart. The size of each bubble corresponds to the number of customers served by the utility, while the center of the bubble (against the y-axis) shows the per household water cost.

Consolidation Into Three Sub-Regional Utilities

Consolidating water utilities into three sub-regional entities (with some systems in outlying areas remaining independent) would reduce the magnitude of the cost increase and would eliminate the high rates that some customers would pay on a stand-alone basis. Figure 4.3 shows that moving water utilities toward a common ownership and management structure—even without physical interconnection or consolidation—could reduce water costs in the Lehigh Valley by more than $16 million in 2020. This represents a 17% reduction in household water

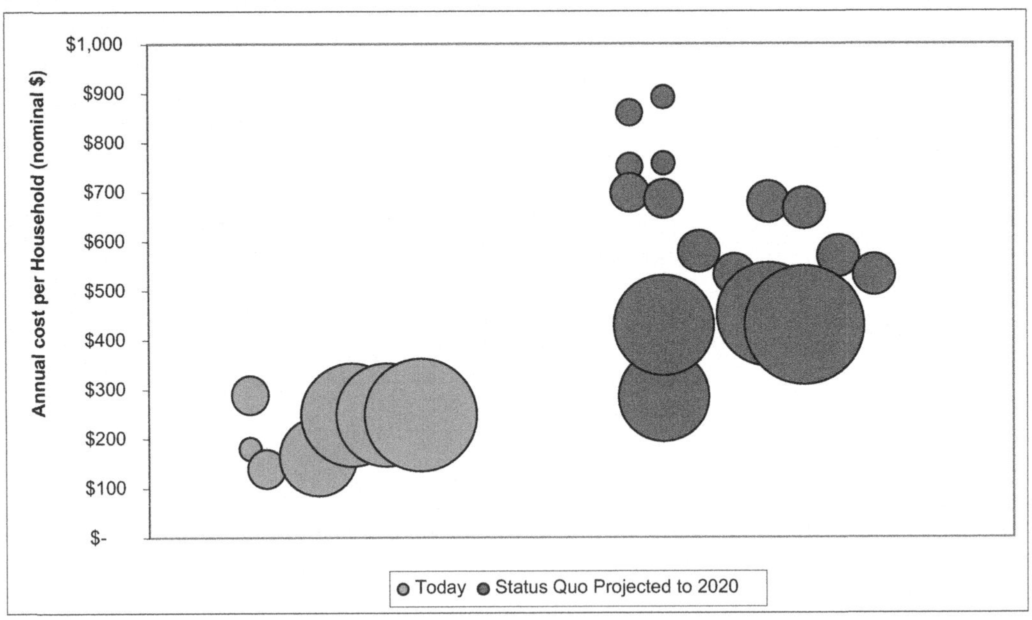

Figure 4.2 Water utility size and typical household cost—present and status quo projected to 2020

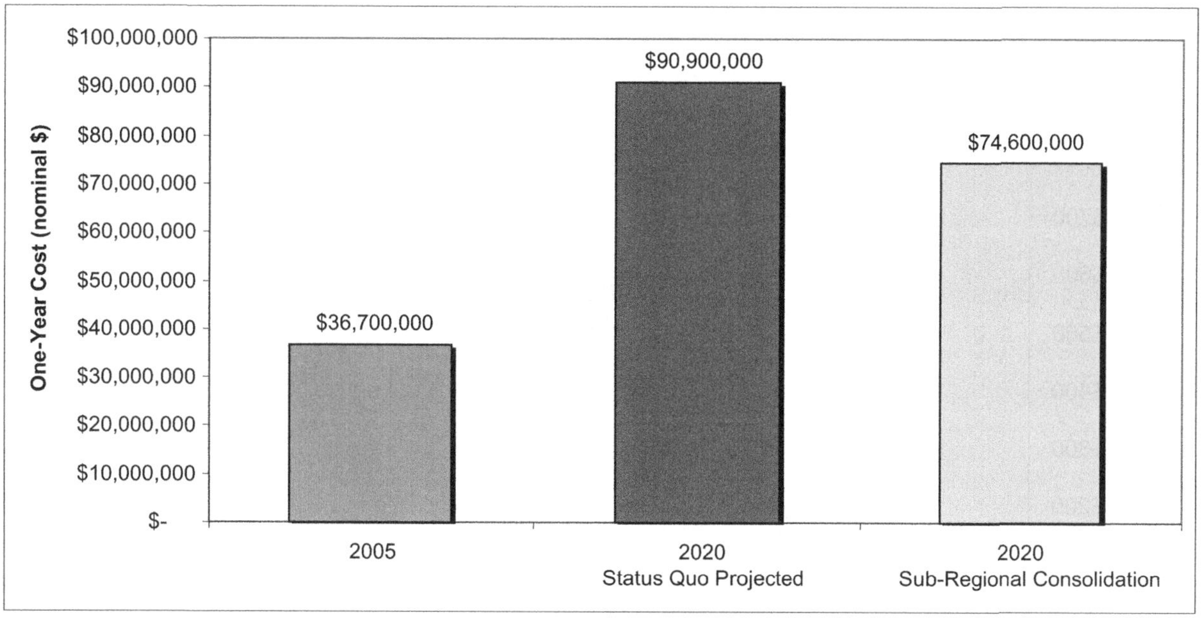

Figure 4.3 Total residential water costs in 2005 and 2020 status quo and sub-regional projected

costs compared to the status quo. It should be emphasized that this savings is for one year only; similar savings would accrue in later (and some earlier) years. Further, similar savings would be achieved for nonresidential customers.

While there are many factors that affect the level of savings, there are two primary drivers of this benefit. First, the efficiency of capital planning is greatly improved with a larger organization. Second, the risk of a catastrophic event is spread over a larger customer base ("pooling" the risk) to minimize the cost impact on each customer.

To put the savings in perspective, there will be approximately 179,000 residential customers in the portions of the Lehigh Valley served by public water suppliers in 2020. A savings of more than $16 million per year is equivalent to approximately $89 per household per year.

Figure 4.4 shows that a sub-regional approach eliminates most of the dispersion in water costs throughout the Lehigh Valley. Residential customers of the three major utilities would have rates ranging from $403 to $420 per year, with an average of $409. A few smaller systems in the northern portion of the Lehigh Valley (serving approximately 2,000 customers) would face water costs that are approximately 50% higher ($630 per year) than the more urbanized portions of the Lehigh Valley.

Consolidation Into One Regional Utility

The next step in the analysis is to model the effect of bringing all public water suppliers in the Lehigh Valley under common ownership and management. Figure 4.5 shows that costs to the Lehigh Valley's households would decline by an additional $6 million per year (more than $33 per household annually) under this ownership and management structure. The typical household water cost would be approximately $380 per year throughout the Lehigh Valley—a reduction of more than 25% compared to the average water cost under the status quo scenario.

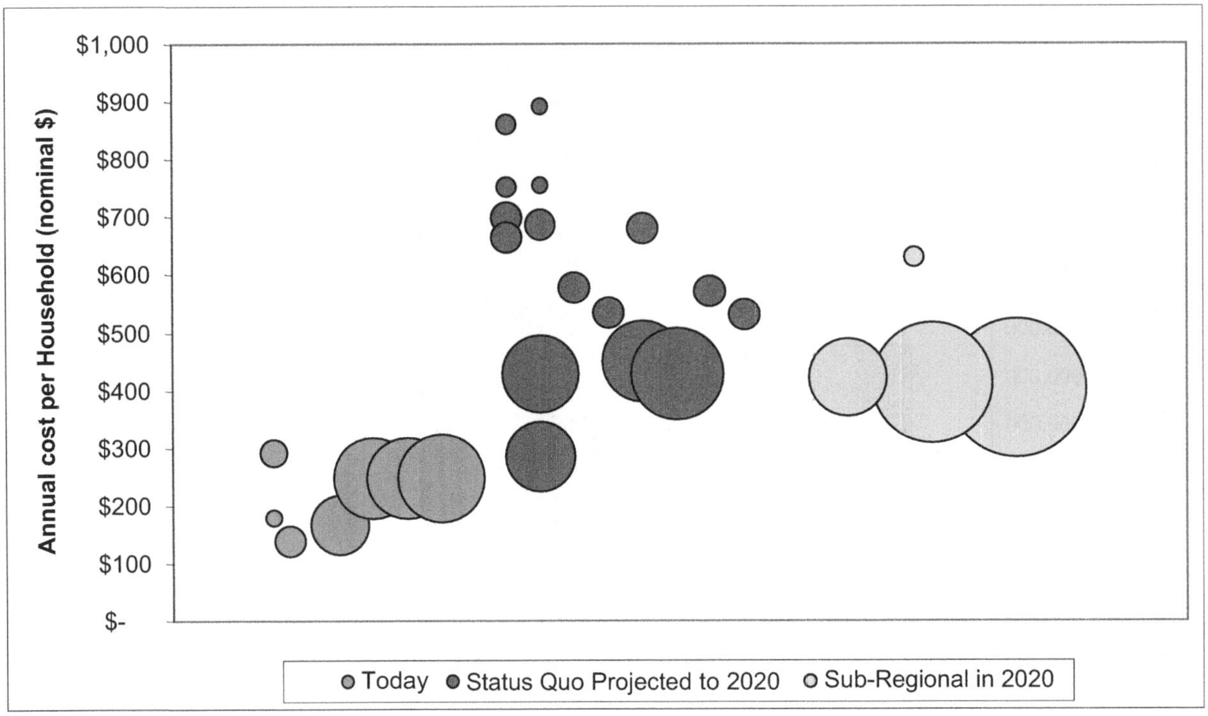

Figure 4.4 Water utility size and typical household cost—present, status quo, and sub-regional projected to 2020

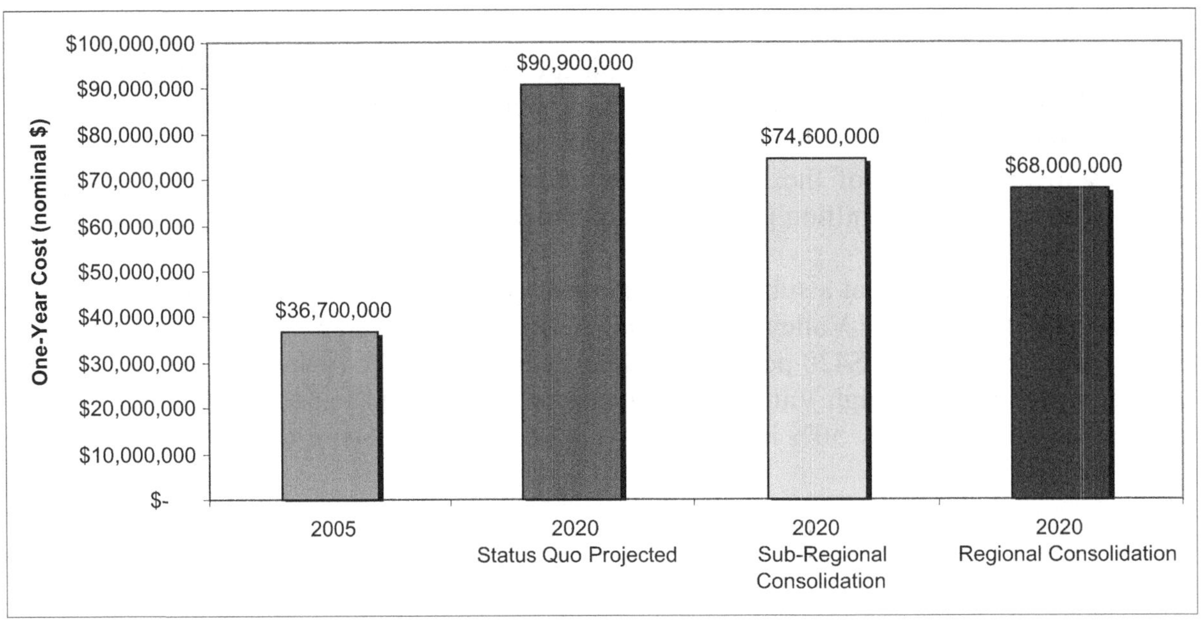

Figure 4.5 Total residential water costs in 2005 and 2020 status quo, sub-regional, and regional projected

The additional savings when compared to the sub-regional scenario are primarily achieved through further improvements in capital planning, additional risk reduction through a combination of enhanced planning and further risk pooling, and increased efficiency in operations (economies of scale).

Testing the Comparability of Model Results

To test the reasonableness of the projections developed using the water cost model, the results were compared to the general relationship between changing the size of a water utility (measured by million gallons produced) and the expected change in cost. Shih and colleagues derived the following equation to represent that relationship (Shih et al. 2006):

$$1-(\Delta\text{volume})^{-0.47} = \%\ \text{change cost} \qquad (4.1)$$

The results of the model for the Cities of Allentown and Bethlehem, on a stand-alone basis (status quo projected) and consolidated basis, were compared to the results predicted by the Shih model.

On a stand-alone basis, the model projects a cost of $501 per household (60,000 gallons) for Allentown. The model also shows water production of 5,698 million gallons in Allentown in 2020. As part of a sub-regional entity, Allentown's production would increase to 10,146 million gallons in 2020. The model shows this sub-regional option would result in a cost per household of $425 in 2020, a reduction in cost of 15.2%. Using the increase in size for Allentown in the Shih equation would estimate the change in cost to be 23.8%. The Shih equation, therefore, would predict a somewhat larger cost savings than the sub-regional model projects for Allentown.

For Bethlehem, the model projects a stand-alone cost of $478 and a sub-regional cost of $427, or a reduction of 10.7%. Water production for Bethlehem would increase from 6,115 million gallons on a stand-alone basis to 7,826 million gallons as a sub-regional entity. The Shih equation would predict this change in size to result in a change in cost of 10.7%. Thus, the Shih equation and the sub-regional model project similar cost savings for Bethlehem becoming part of a sub-regional entity.

Conducting the same analysis for Allentown and Bethlehem becoming part of a single regional entity leads to a more divergent result. The regional model for Lehigh Valley predicts cost savings in Allentown and Bethlehem ranging between 23% and 26%, while the Shih equation suggests substantially higher cost savings in the 48% to 49% range.

The reasons for the divergence in results at the regional level were not fully investigated. It could be that the regional model for Lehigh Valley has failed to capture some cost savings that could be achieved. Or it could be that the Shih model was not meant to be scaled up to the magnitude that would be involved in a Lehigh Valley-wide consolidation (Allentown and Bethlehem would become part of a regional entity that is approximately four times the size of each utility separately, as measured by production).

In any event, it appears that the water models developed for the Lehigh Valley reasonably reflect the magnitude of cost savings that may be achievable through sub-regional consolidation. It also appears that the Lehigh Valley models may understate the potential savings achievable from total consolidation throughout the Lehigh Valley.

WASTEWATER

Status Quo Projected to 2020

In 2005, the typical household in the Lehigh Valley paid a wastewater bill of less than $250 per year, with bills ranging from $190 to $440 per year (based on discharge of 60,000 gallons per year). In total, Lehigh Valley wastewater suppliers serve approximately 243,000 residential customer accounts. At present, it is estimated that the total annual wastewater bill for those residential customers is on the order of $56 million per year.

Using the wastewater cost model developed for this project, and assuming no structural changes in the organization and operations of the Lehigh Valley's wastewater utilities, it is estimated that these costs will increase significantly by 2020. Most customers are likely to see their wastewater bills double, and in some cases triple, by 2020.

In particular, it is estimated that the average wastewater cost will increase from its current Lehigh Valley-wide average of about $230 per year to approximately $480 per year in 2020. This would result in a total residential wastewater cost of approximately $130 million in 2020, an increase of nearly 130% over existing costs, as shown in Figure 4.6.

Today more than 96% of residential wastewater customers pay between $190 and $250 per year—a range of about 18% below average to 9% above average. Fewer than 8,000 households served by small wastewater utilities currently pay costs above $400 per year. By 2020, costs for all utilities will increase significantly, but the steepest increases will affect customers of the smallest utilities, as shown in Figure 4.7.

As was the case for water utilities, there are many factors that influence the projection of costs in 2020. There are, however, three main drivers for the wastewater cost increases: (1) inefficiency in capital planning and investment; (2) inability to spread risk; and (3) general inflation.

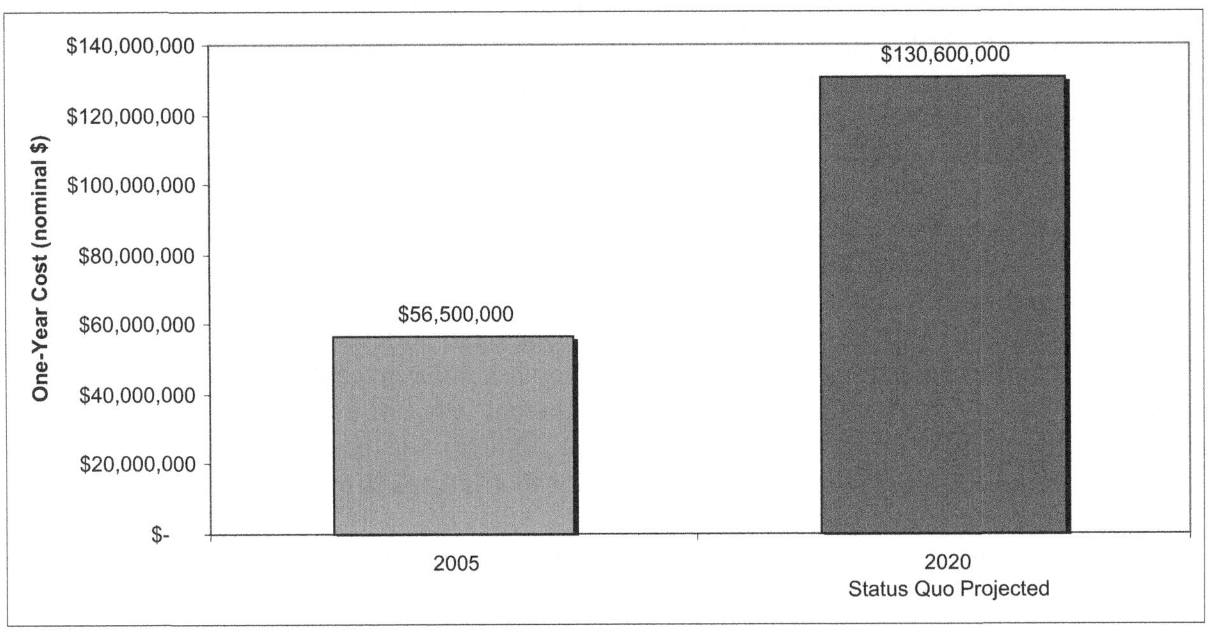

Figure 4.6 Total residential wastewater costs in 2005 and 2020 status quo projected

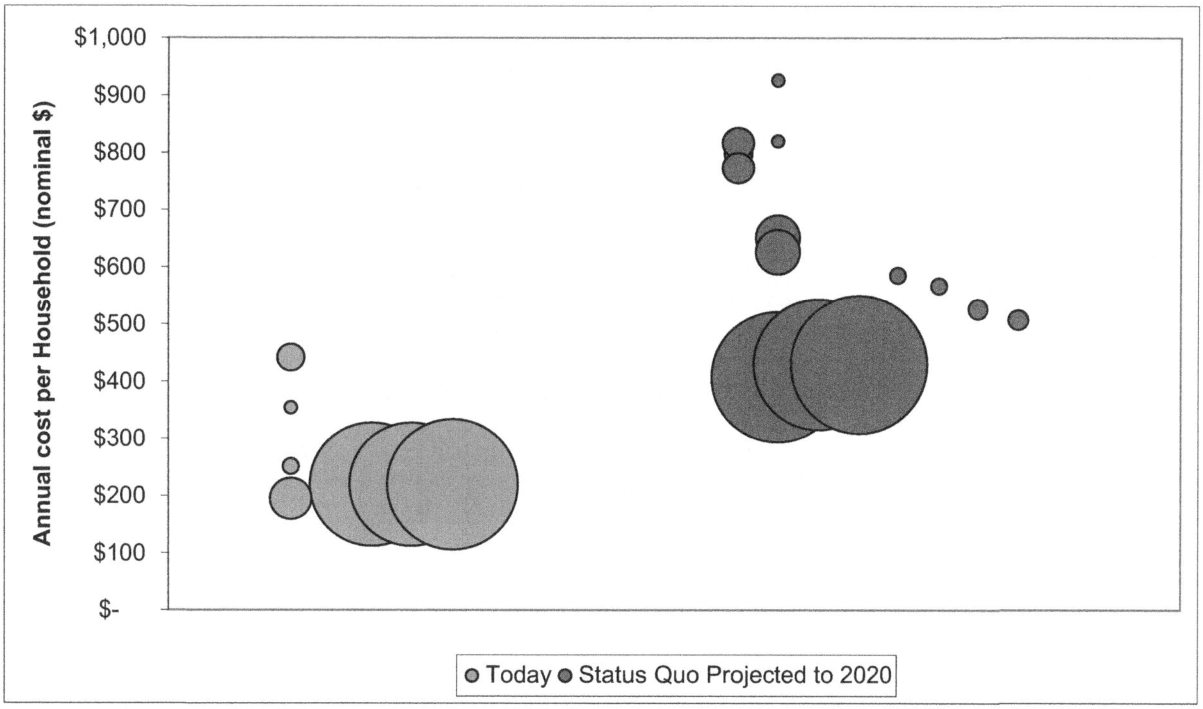

Figure 4.7 Wastewater utility size and typical household cost—present and status quo projected to 2020

Consolidation Into Sub-Regional Utilities

Consolidating wastewater utilities into three large sub-regional entities (with utilities in outlying areas of the Valley remaining independent) would reduce the magnitude of the cost increase (especially for smaller utilities) and would eliminate the extremely high rates that some customers would pay on a stand-alone basis. Figure 4.8 shows that moving wastewater utilities toward a common ownership and management structure—even without physical interconnection or consolidation—could reduce wastewater costs in the Lehigh Valley by more than $25 million in 2020. This represents nearly a 20% reduction in household wastewater costs compared to the status quo. It should be emphasized that this savings is for one year only; similar savings would accrue in later (and some earlier) years. Further, similar savings would be achieved for nonresidential customers.

While there are many factors that affect the level of savings, two factors provide a major portion of the benefit. First, the efficiency of capital planning is greatly improved with a larger organization. Second, the risk of a catastrophic event is spread over a larger customer base.

To put the savings in perspective, there will be approximately 272,000 residential customers in the portions of the Lehigh Valley served by centralized wastewater systems in 2020. Savings of more than $25 million per year represent a savings of approximately $92 per household per year.

Figure 4.9 shows that a sub-regional approach eliminates the spikes in wastewater costs that smaller systems would have experienced in 2020. Typical residential bills would range between $350 and $500, with most customers paying $410 per year or less.

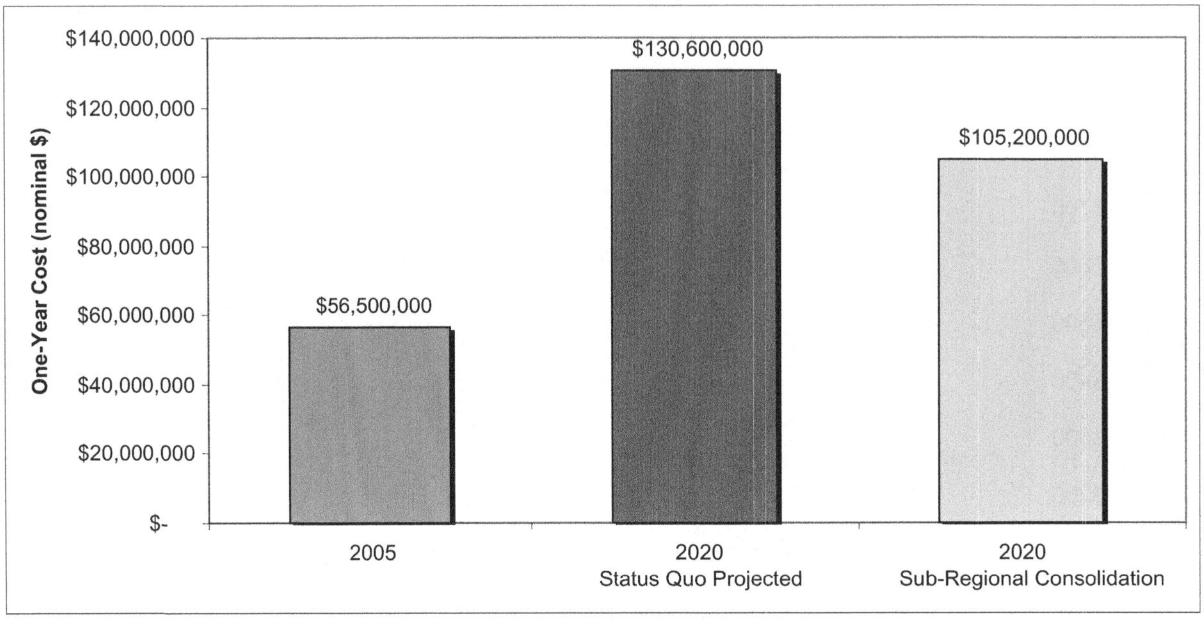

Figure 4.8 Total residential wastewater costs in 2005 and 2020 status quo and sub-regional projected

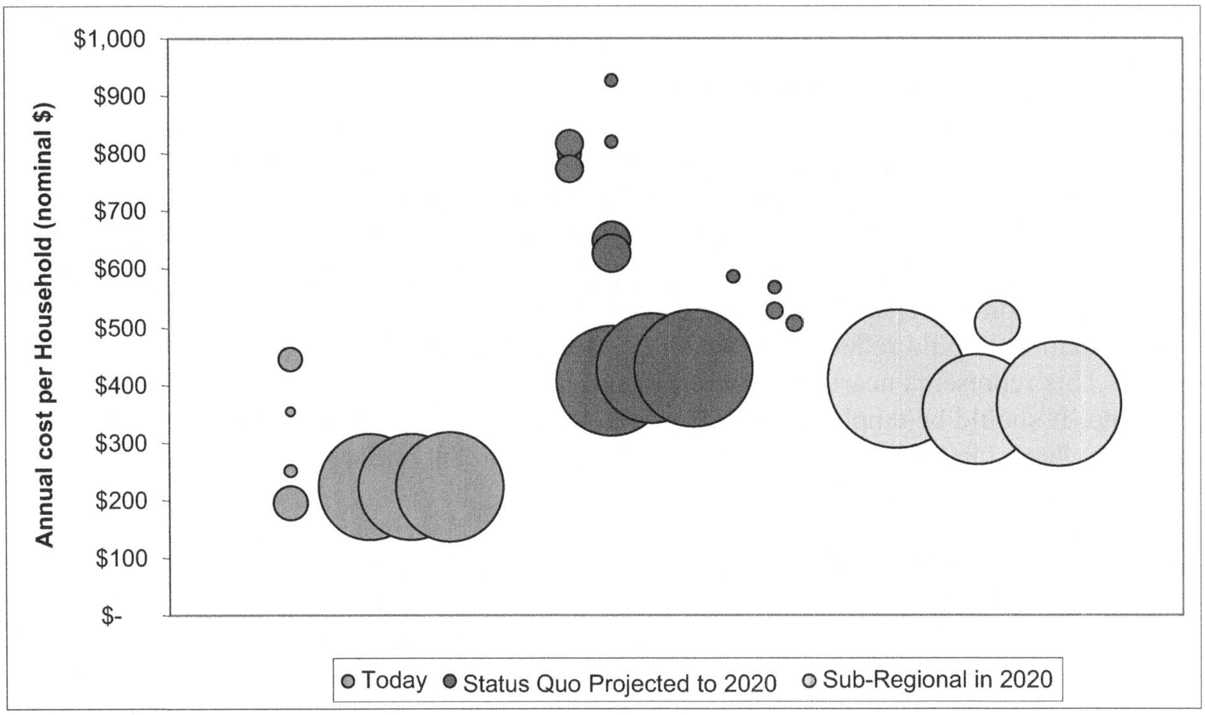

Figure 4.9 Wastewater utility size and typical household cost—present, status quo, and sub-regional projected to 2020

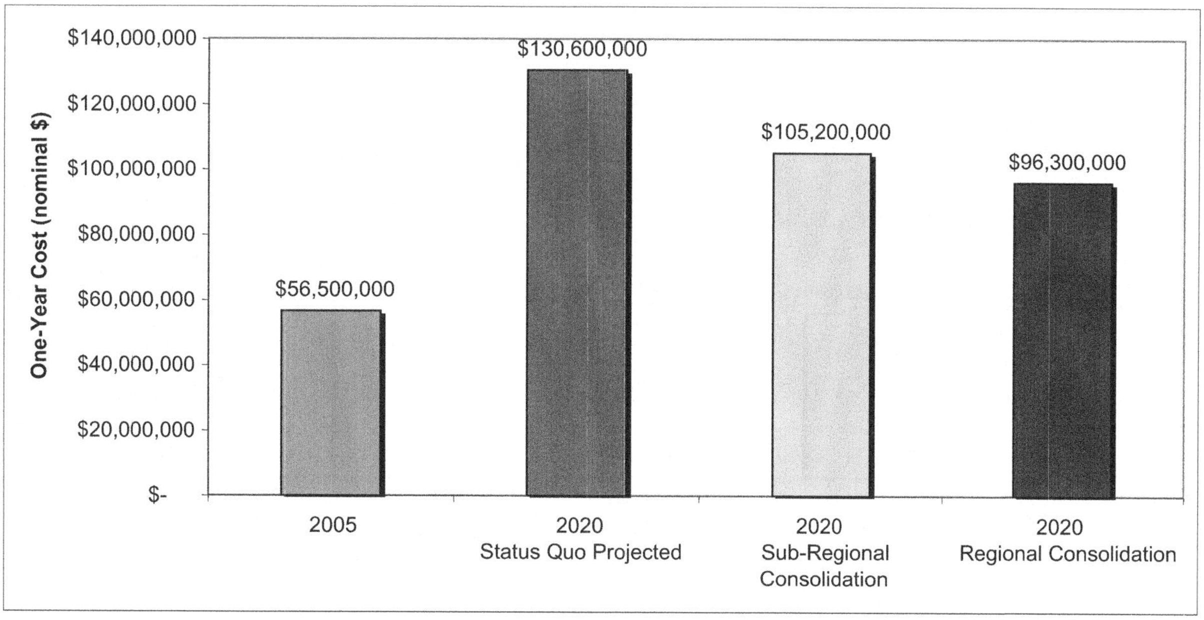

Figure 4.10 Total residential wastewater costs in 2005 and 2020 status quo, sub-regional, and regional projected

Consolidation Into One Regional Utility

The final step in the wastewater analysis is to model the effect of bringing all public wastewater utilities in the Lehigh Valley under common ownership and management. Figure 4.10 shows that the cost to the Lehigh Valley's households would decline by an additional $9 million per year (approximately $33 per household annually) under this ownership and management structure. The typical household wastewater cost would be approximately $350 per year throughout the Lehigh Valley—a reduction of more than 26% compared to the average wastewater cost under the status quo scenario. The additional savings when compared to the sub-regional scenario are primarily achieved through further improvements in capital planning, additional risk reduction through a combination of enhanced planning and further risk pooling, and increased efficiency in operations (economies of scale).

Overall Impact on Water and Wastewater Costs

In summary, the Lehigh Valley could achieve significant savings in water and wastewater costs through the consolidation of water and wastewater suppliers. Figure 4.11 shows that the combined increase in water and wastewater costs by 2020 is projected to be more than $128 million on an annual basis. Sub-regional consolidation could reduce the magnitude of this cost increase by approximately $42 million, or by 32%. Further consolidation into a single regional water and wastewater provider could further reduce costs by an additional $15 million, or $57 million in total. Thus, regional consolidation is projected to avoid more than 44% of the projected increase in water and wastewater costs between 2005 and 2020.

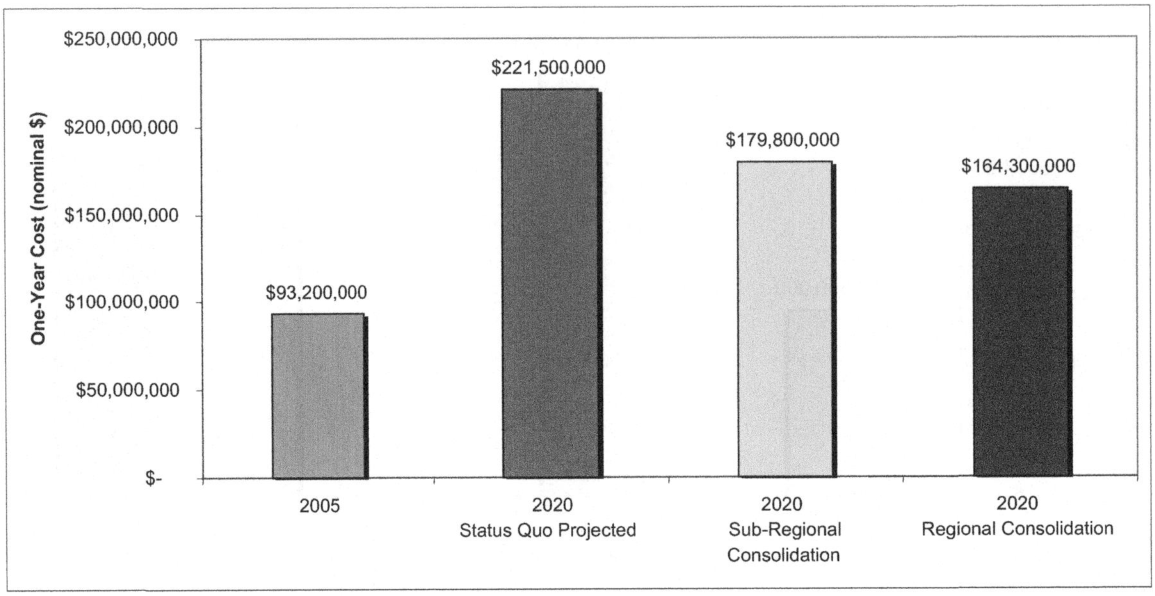

Figure 4.11 Total residential water and wastewater costs in 2005 and 2020 status quo, sub-regional, and regional projected

CHAPTER 5
FINDINGS AND CONCLUSIONS

This research project has employed a business planning approach to evaluate the long-term prospects of 40 publicly owned utilities providing water and wastewater services in the Lehigh Valley of Pennsylvania. This involved four analytical steps:

- The first step in implementing the research approach was the establishment of a baseline understanding of utility performance based on 2005 operating and financial reports and individual utility interviews.
- The second step was to forecast the prospective condition of these existing utilities in the year 2020. These "status quo" or "no action" scenarios take account of the increasing challenges of the business environment that are likely to impact the water and wastewater industry over this 15-year period. This analysis modeled no changes from the status quo in terms of the form and scale of organization of the 40 utilities in the Lehigh Valley.
- The third step was to develop an alternative set of forecasts to simulate the potential of expanded regional collaboration at a larger scale to offset the advancing challenges of the projection period.
- Finally, the difference between the regional consolidation scenarios and the "status quo scenarios" was computed in terms of the difference in the annual household cost of water and wastewater service. This difference is interpreted as the potential benefit of expanded regional collaboration. When multiplied by the number of households, it provides an estimate of the total annual savings to the region as a whole.

FINDINGS

Under the status quo scenario, the average household in the Lehigh Valley, paying about \$220 per year for water service in 2005, will pay about \$510 per year in 2020. The same average household pays about \$230 per year for wastewater service in 2005, which will escalate to about \$480 per year in 2020 under the status quo scenario. The total water and wastewater bill for the average household is projected to more than double from \$450 per year in 2005 to \$990 per year in 2020. Computing results for the average household masks considerable variation in the results. At the extremes, some small communities could see a tripling or quadrupling of the household cost by the year 2020 with combined water and wastewater bills approaching \$1,800 per year or more in some communities.

Results for the sub-regional approach to consolidation that would create three large water and wastewater supply entities in the Lehigh Valley (surrounding Allentown, Bethlehem, and Easton) indicate that 2020 water bills for the average household would be about \$820 per year, divided equally between water and wastewater. By comparison to the status quo scenario, this represents an average household savings of more than \$170 per year. Regionwide, this amounts to a total savings of \$41 million per year by 2020. It is important to stress that these savings would accrue not just once, but every year.

Results for the total consolidation scenario where all water and wastewater suppliers would be owned and managed by a single regional entity indicate that 2020 water bills for the average household would be about \$380 per year and wastewater bills would be about \$350 per year. By comparison to the status quo scenario, this represents an average household savings of $260 per year. Regionwide, this amounts to a total savings of $56 million per year by 2020. Again, these savings would accrue not just once, but every year.

CONCLUSIONS

Water and wastewater costs are likely to increase faster than the rate of inflation in the Lehigh Valley during the next dozen years. The magnitude of these cost increases is not inevitable. There is currently a limited window of opportunity during which public water and wastewater utilities in the Lehigh Valley could save tens of millions of dollars each year through enhanced regional collaboration. Significant savings can be achieved through improved planning, financial management, risk reduction, facilities planning, infrastructure management, and workforce management. If action is not taken soon, however, many of these opportunities will be lost.

There are opportunities to manage and mitigate the risks and long-term consequences associated with population growth. But those opportunities will diminish over time and may no longer exist after the growth has occurred. Given the level of projected population growth within the next dozen years, the delay of more than a few years will result in a significant lost opportunity.

Allentown and Bethlehem will need to substantially upgrade their water and/or wastewater treatment plants within the next few years. If regional plans are not developed soon, the opportunity to enhance the utilization of those regional facilities may be missed.

Addressing the aging water and wastewater pipe infrastructure is a long-range challenge. The key to meeting that challenge is to undertake a comprehensive data collection, analysis, and planning process. But that process must begin within the next few years, or the costs of dealing with the failure to act will soon overwhelm utility budgets and other resources.

The looming retirement of the "baby boom" generation creates a unique opportunity to address utility workforce issues throughout the Lehigh Valley while minimizing the number of people (if any) who would lose their jobs. But failing to design and implement these plans within the next few years could miss this opportunity.

In summary, a number of factors have converged to make this an ideal time to consider an enhanced level of consolidation and regional cooperation in the Lehigh Valley. The authors estimate that such a regional effort could save the average household as much as $260 per year in 2020—amounting to $56 million in annual savings for the region as a whole.

CHAPTER 6
RECOMMENDATIONS

This research project was devised in part to follow up on recommendations of prior AwwaRF research (Raucher et al. 2006) which had called for a new paradigm to evaluate regional collaboration. It was observed that traditional approaches of weighing pros and cons provide an unbalanced view because potential benefits are not made more tangible. When the potential economic benefits are vaguely defined, they are often too amorphous to overcome the inertia of the status quo. When everything seems fine in the short-term, there is a tendency to not fix what does not appear broken. Thus, the scales used in comparison of pros and cons are not balanced. The proposed new paradigm is called "benefits and issues." The intent is to first quantify the potential benefits of regional collaboration and then convene a process to discuss the benefit estimates and to identify and discuss the issues that would need to be addressed in order to devise a means of capturing those benefits.

This study of 40 utilities in the Lehigh Valley of Pennsylvania has demonstrated that there are substantial potential benefits to be gained from enhanced regional collaboration. The logical next step is to convene a process to discuss the means of capturing these benefits and to identify potential issues that must be resolved. The Pennsylvania Municipality Authorities Act provides a very flexible framework within which new institutions could be created at either the sub-regional (e.g., Easton, Easton Suburban) or regional level to consolidate utility assets and/or functions to deliver more sustainable and affordable water and wastewater services to the benefit of all jurisdictions.

Consolidation raises a range of legitimate issues that are of concern to everyone. This is where the range of flexibility provided within the Pennsylvania Municipality Authorities Act is extremely valuable. It is important to stress that, under the Act, the starting point for creation of a *new* authority, at either the sub-regional or regional level, would be a "clean sheet of paper." The parties wishing to design a new collaboration could design it from scratch to suit their needs and ameliorate any concerns they have. They can design features to prevent them from losing control through the charter and through their representation on the Board. They can address concerns over service continuity, quality, and cost control by stipulating performance requirements and guarantees. They can provide hands-on management of workforce transition issues. They can work through the due diligence details of asset transfers to achieve consolidation in a transparent process that will provide a clear basis for making the new arrangements more equitable for all concerned. In short, every issue that needs to be resolved in order to capture the benefits the authors have identified can be addressed on a "clean sheet of paper." Of course, no one has to sign on until they are fully satisfied that their concerns have been addressed. With a likely range of benefits now identified, the authors believe it is timely to convene a process to discuss the issues involved in moving forward to capture those benefits.

There is an opportunity to meet the challenges of the future with alternative approaches to regional collaboration that can increase the Lehigh Valley's ability to respond to change. But much of the benefit computed here would result from responding to change as it is happening, not responding after the fact. In order to capture these benefits, it is necessary to implement broader regional collaboration relatively soon. The most challenging cost influences that are bearing down on the utilities in the Lehigh Valley are visible today and may cause many individual utilities to make irreversible commitments that will negate the longer-term opportunity.

Lost opportunity is not the only motivation for early action to promote broader collaboration. It is clear that in the current configuration, individual jurisdictions are so limited in their ability to cope with change and mitigate risk that many of their available responses result in the acceptance of extraordinary levels of financial risk for public enterprises. At some point, this is not sustainable. The authors project that some utilities in the Lehigh Valley could see a tripling or even quadrupling of water and wastewater bills by 2020 unless they take action soon to manage the inherent risks of the business.

CHAPTER 7
RESEARCH IMPLICATIONS

The focal point of the authors' methodology for examining the potential benefits of enhanced regional collaboration is the baseline projection of the status quo into the future. What will happen if you "do nothing" to adapt water and wastewater utilities to meet changing operating conditions? What does that look like 15 years hence?

This approach raises a set of questions that is not routinely considered in the day-to-day affairs of a typical utility. The normal perspective is short-term; and, in the short-term, things generally look good. The financial reports reviewed in this study showed all the utilities to be solvent. Even when there is occasional trouble of the type that was described to the authors by many utilities, it can be viewed as a bump in the road. In essence, when there is no long-term view of how things are changing, there is no indication of a need for change.

If a utility does not see change coming, it cannot adapt. A sustainable utility business is one that knows what is coming and understands its options for adaptation. It is imperative to ask questions about the longer-term, like those the authors have asked:

- What risks are entailed in current arrangements for the provision of treatment capacity, whether as owner or wholesale customer? How stable are these arrangements and relationships? How dependent are they on growth projections and revenues from connection charges? How vulnerable are utility finances to adversities that can affect facilities and operations—are there sufficient cash reserves to restore service in the event of a major failure in any part of the system?
- What provision is being made for rehabilitation and replacement of pipe infrastructure? Is the strategy entirely reactive or does it include a planning process to target needs? Does it include a commitment to actually ramp up reinvestment in the infrastructure?
- What is the plan for workforce succession in the wake of baby boom retirements? How will the utility recruit the next generation in a much tougher labor market? Is the current staffing level and deployment efficient in light of how the labor market is tightening?

The authors' modeling approach allows explicit assessment of the effects of these and other cost factors on the long-term finances of the utility. In the status quo scenario, the authors assumed that utilities would continue trying to meet these challenges within existing arrangements. In the authors' alternative scenarios, the authors modified these assumptions based on the authors' assessment of the cost savings that are possible by pooling risks and operating at a larger scale.

The authors' spreadsheet model is provided as a tool accompanying this report. It is pre-populated with the generic financial profiles of different types and sizes of utilities that the authors developed in this study. However, it allows the user to modify the inputs based on their own financial profile as reflected in their own financial report. It then allows the user to do as the authors have done and make varying assumptions to project the status quo and alternatives into the future. To create an alternative future, start with the model for a larger entity and see how the financial outcomes change if another utility's customers and facilities are added to that picture. More instructions are provided with the tool.

One of the side objectives of this research project was to accomplish a comprehensive study of regional collaboration on a relatively modest budget. The authors' work entailed conceptualization and construction of the spreadsheet tool as well as a complete application of it in the Lehigh Valley as a demonstration of the approach. The authors believe that the business planning approach to comparison of alternative organizational strategies is universally applicable to utilities in any region.

The resulting spreadsheet model has transferable applicability to other regions as well. Current operating and financial reports and a sense of future trends is needed to customize it to another region. However, in its present form, the spreadsheet model is such that an experienced business/financial analyst would be more comfortable adapting it to local circumstances than a user without that type of background.

Given the broad applicability of the concepts employed, and their fundamental connection to important themes of water system viability and sustainability, the authors believe that the development of a more robust and broader user-friendly analytical tool built upon this foundation would merit consideration by AwwaRF as a logical extension of this research. Additional demonstrations of a more advanced tool as Beta tests would also add more empirical quantifications of benefits to the literature which could greatly improve regionalization efforts.

APPENDIX A
DETAILED MODELING INPUTS AND ASSUMPTIONS

This appendix briefly explains the basis and source for each of the major assumptions and inputs used for projecting costs to 2020 in the water and wastewater cost models. As used in this appendix, a "real" cost change refers to the cost above the general rate of inflation. For example, if the general inflation rate is 2.5% per year, a real cost change of 1% per year would equal an annual change of 3.5%.

Capital Replacement, Excluding Water Mains. $200 per year in 2005 dollars for each million gallons of water produced annually. For example, a system that produces 1 million gallons per day would produce 365 million gallons per year, so this assumption would result in a capital replacement budget of $70,800 (365 × $200) in 2005 dollars. This assumption was based on the ratio of the City of Allentown's capital improvement plan to its annual water production, and was verified by comparing the capital budgets and production levels of other water utilities in the Lehigh Valley.

Connection Fee per Customer, Wastewater. $4,000 in 2005 dollars, based on Lehigh County Authority actual charges.

Connection Fee per Customer, Water. $2,000 in 2005 dollars, based on Lehigh County Authority actual charges.

Construction Cost Escalator. 1.4% annual increase in the real cost of construction, based on the average annual change in construction costs above general inflation in the Philadelphia region from year-end 2004 to year-end 2006. Cost Indexes by City, Engineering News-Record (December issue of each year).

Debt Service Coverage. 1.50 times, based on median for A-rated municipal water and wastewater utilities. Fitch Ratings, Ltd., 2007 Median Ratios for Water and Sewer Revenue Bonds—Retail Systems (January 16, 2007).

Electricity Cost Escalator. 6.8% reduction in the real cost of electricity in 2020 compared to 2005, based on the projected cost per kilowatt-hour for commercial customers in 2020 compared to the average cost per kilowatt-hour for commercial customers in 2005. Energy Information Administration, US Department of Energy, Annual Energy Outlook 2007, Report No. DOE/EIA-0383(2007), Table 8 (Electricity Supply, Disposition, Prices, and Emissions) (December 2006).

General Inflation. 2.56% annual level of inflation, based on the average annual change in the Consumer Price Index in the Philadelphia region from December 1996 to December 2006. Bureau of Labor Statistics, U.S. Department of Labor, http://www.bls.gov/cpi/.

Interest Rate, Municipal Bonds. 4.78% interest rate, based on the projected interest rate on AA-rated utility bonds in 2010, multiplied by 65% to recognize the tax-exempt nature of municipal bond debt. Energy Information Administration, U.S. Department of Energy, Annual Energy Outlook 2007, Report No. DOE/EIA-0383(2007), Table 19 (Macroeconomic Indicators) (December 2006)

Labor Costs. 1% annual increase in the real cost of labor, based on the average annual change in wages above general inflation in the water and sewer industry in Pennsylvania from 2001 through 2005. Bureau of Labor Statistics, U.S. Department of Labor, Quarterly Census of Employment and Wages, http://www.bls.gov/cew/.

Operating and Maintenance Expenses (Non-Labor, Non-Energy). 1% annual increase in the real cost of operations and maintenance. The estimate is based on the change in labor costs and general observations of the magnitude of cost changes in the participating utilities over the past two years.

Operating Ratio. 1.30, based on John E. Cromwell III, et al., Business planning for small system capacity development, JAWWA 89:1:47–57 (January 1997), and Fitch Ratings, Ltd., 2007 Median Ratios for Water and Sewer Revenue Bonds—Retail Systems (January 16, 2007).

Population Projections. Prepared by Lehigh Valley Planning Commission. Lehigh Valley Planning Commission, Municipal Population Forecasts, Lehigh & Northampton Counties (June 4, 2007), http://www.lvpc.org/MunPopulationForecasts.pdf.

Wastewater Treatment Plant Upgrade, Large Utility, Low Projected Growth. $30,000,000 in 2005 dollars, based on estimate provided by City of Allentown.

Wastewater Treatment Plant Upgrade, Large Utility, Moderate Projected Growth. $53,000,000 in 2005 dollars, based on estimate provided by City of Bethlehem.

Wastewater Treatment Plant Upgrade, Medium Utility. $3,350,000 in 2005 dollars, based on equations developed by the authors from the USEPA 1996 Wastewater Needs Survey for treatment and interceptor costs based on population served, inflated from 1996 to 2005, designed to serve projected population in 2030.

Wastewater Treatment Plant Upgrade, Small Utility. $1,560,000 in 2005 dollars, based on equations developed by the authors from the USEPA 1996 Wastewater Needs Survey for treatment and interceptor costs based on population served, inflated from 1996 to 2005, designed to serve projected population in 2030.

Water Main, Installed Cost per Foot. $200 in 2005 dollars, based on City of Allentown's typical actual cost.

Water Main Replacement Rate. 0.70% per year for a stand-alone utility; 0.35% per year for sub-regional organizations; 0.30% per year for a regional organization, based on analysis of pipe replacement data for a sample of water utilities (stand-alone), the average for the 20 largest utilities in the sample (sub-regional), and the best performers in the sample (regional) (Cromwell, Speranza, and Reynolds 2001).

Water Treatment Plant Capital Improvements. $25,000,000 in 2005 dollars, based on actual cost expended by City of Allentown and City of Bethlehem in mid-1990s for treatment plant upgrades.

Water Well Construction Cost. $2,000,000 in 2005 dollars, based on information provided through interviews with various water utilities in the Lehigh Valley.

APPENDIX B
INSTRUCTIONS FOR USING THE WATER MODEL

The water and wastewater models are available on a CD-ROM that accompanies this report. The format, and most inputs, for the two models are essentially the same. This appendix, therefore, will focus on the water model, but the instructions and descriptions are equally applicable to the wastewater model.

The model contains 14 worksheets. There are worksheets containing assumptions, data inputs, and outputs. For ease of understanding, the tab for each worksheet is color-coded (tab colors display in versions of Microsoft Excel published beginning in 2002):

- Assumptions tabs are yellow
- Input tabs are green
- Output tabs are red

ASSUMPTION WORKSHEETS (YELLOW TABS)

The key assumptions used in the model are identified on the **Assumptions** worksheet. All values on this worksheet are named variables. Entries on this worksheet are color coded by their type or source:

- Values with a white (or clear) background are input on this worksheet.
- Values with a blue background contain drop-down boxes that allow the user to select among two or three options.
- Values with a gray background contain formulas that calculate the value from other information on the Assumptions worksheet. *These cells should not be changed.*
- Values with a pink background contain information that is input or calculated on one of the input worksheets. *These cells should not be changed.*

Values That Can be Changed (White or Clear Background)

- Connection fee revenue per new customer (CONNECT): Assumed value is $2,000 for water ($4,000 for wastewater).
- Cap on debt as % of net plant (DEBT_CAP): This represents the upper limit on the percentage of utility plant that can be financed with debt. The assumed value is 100%.
- Required ratio of revenues to expenses (REV_EXP): The assumed minimum operating ratio to maintain financial viability is 1.30, as discussed in chapter 3.
- Debt service coverage requirement (DEBT_COVER): The assumed minimum debt service coverage to maintain financial viability is 1.50, as discussed in chapter 3.
- Operations and maintenance efficiency from sub-regions (OM_LEVEL2): This value assumes that a certain level of efficiency in operating and maintenance expenses can be obtained through sub-regional consolidation. The assumed level of cost savings is 20%, so the assumed value for this variable is 0.80.
- Plant expansion/upgrade year (PLANT_YEAR, PLANT_YEAR_GROW, PLANT_YEAR_LOW, PLANT_YEAR_ALLEN, PLANT_YEAR_BETH): These

values reflect assumptions about the year in which there will be a significant investment in upgrading or adding treatment and/or production plant. The assumed values are 2010 and 2015 for high-growth utilities, 2015 for low-growth utilities, and 2020 for Allentown and Bethlehem.

Values That Can be Selected Using a Drop-Down Box (Blue Background)

- Electricity cost option (ELEC_OPTION): There is a choice of three electricity price forecasts: the forecast prepared by the Energy Information Administration in 2007 (the ELECTRIC variable), an assumption of no real increase in electricity costs (that is, costs will increase only with general inflation), or 1% real increase in costs (that is, costs will increase by one percentage point more than inflation in each year). Changing ELEC_OPTION automatically changes the electricity cost escalator to use (ELEC_USE).
- Construction cost option (CONST_OPTION): There is a choice of three construction cost escalation forecasts: a forecast based on the actual change in Philadelphia region construction costs from 2004 through 2006 (the CONSTRUCTION variable), an assumption of no real increase in construction costs, or a 1% real increase in costs. Changing CONST_OPTION automatically changes the construction cost escalator to use (CONST_USE).
- Consolidation option (CONSOLIDATION): Use this drop-down box to control the scenario being evaluated: No Consolidation (that is, the status quo projected), Sub-regions (the creation of four utilities in the region), or One Region (the creation of a single regional utility)
- Risk option (RISK): Select "TRUE" or "FALSE" from this drop-down box to select whether the scenario should include the impact of a "risk event" which adds the equivalent of a new well in 2013.

Assumptions about the make-up of each sub-region are contained on the **Subregions** worksheet. This worksheet enables the user to enter the number of water systems of each type (small no growth, small growth, etc.) that will be in each of the four sub-regions. The number of systems of each type must equal the specific number shown here (the "Check" column ensures that the totals for each system type are accurate).

INPUT WORKSHEETS (GREEN TABS)

There are five worksheets containing the critical input variables for the model. *Items on these sheets should not be changed* unless the user wishes to update source data with more recent figures (for example, more recent inflation or construction cost data). The input worksheets contain the following types of data:

- Inputs: Data on population, electricity costs, summary of cost escalation data (electricity, construction, labor, other operations and maintenance expenses), interest rate summary, main installation costs, and main replacement factors.
- BaseCase: Detailed financial information for each type of utility.

- InterestRates: The annual forecast of interest rates on AA-rated utility bonds contained in the econometric model used by the Energy Information Administration.
- Wages: Data from the Bureau of Labor Statistics on wages paid in Pennsylvania to people working for water or wastewater utilities.
- Escalation: Construction cost (ENR) and inflation (Consumer Price Index, Philadelphia region) indexes for each year from 1996 through 2006.
- Pipe: Calculation of escalation costs for pipe replacement investments and non-pipe capital investments, using the variables CONST_USE, MAIN_COST, and REPLACE. This worksheet calculates the variables PIPE_CAPITAL and REPLACE_CAPITAL that are used in the model. *This worksheet should not be changed.*

OUTPUT WORKSHEETS (RED TABS)

There are six worksheets containing the outputs from the model.

The **Output** worksheet shows a summary of the information calculated on the other output worksheets. The Output worksheet shows the results of the particular scenario being evaluated (the specific combination of options selected or entered on the Assumptions worksheet).

Important Note: This model does not compare scenarios. It produces the output for one scenario only. In order to compare the results of different scenarios (different sets of assumptions), the information from the relevant portion of the Output worksheet (Status Quo, Sub-regions, or One Region) must be printed, or copied to a new table using Edit/Paste Special/Values, to "freeze" the results of the scenario.

The other output worksheets contain the detailed calculations for each type of utility (Small, Medium, LCA, Large, and DistributionOnly).

None of the information on any of the output worksheets should be changed. All changes should be made on the assumptions worksheets.

REFERENCES

AWWA (American Water Works Association). 2000. *Manual M1: Principles of Water Rates, Fees, and Charges (5th Ed.)*. Denver, Colo.: AWWA.

Cromwell, J.E., S. Rubin, F. Marrocco, and M. Leevan, 1997. Business Planning for Small System Capacity Development. *JAWWA*, 89(1)47–57. January.

Cromwell, J.E., E. Speranza, and H. Reynolds. 2001. Dawn of the Replacement Era: Reinvesting in Drinking Water Infrastructure. Denver, Colo.: American Water Works Association. June.

Cromwell, J.E., E. Speranza, G. Wammock, P. Demit, and H. Reynolds. 2006. Water Infrastructure at a Turning Point: The Road to Sustainable Asset Management. Denver, Colo.: American Water Works Association. June.

EIA (Energy Information Administration). 2006. *Annual Energy Outlook 2007*. Report No. DOE/EIA-0383(2007). U.S. Department of Energy. December.

Fitch. 2007. Median Ratios for Water and Sewer Revenue Bonds—Retail Systems. New York, N.Y.: Fitch Ratings Ltd. January 16.

Katz, B., A. Liu, M. Muro, J. Vey, K. Sommer, and D. Jones. 2003. Back to Prosperity: A Competitive Agenda for Renewing Pennsylvania. Washington, D.C.: The Brookings Institution Center on Urban and Metropolitan Policy.

LVPC (Lehigh Valley Planning Commission). 2007. Municipal Population Forecasts Lehigh & Northampton Counties. Lehigh Valley Planning Commission, Allentown, Pa. Available: <http://www.lvpc.org/MunPopulationForecasts.pdf>. [cited June 4, 2007]

Raucher, R.S., J.E. Cromwell, J. Henderson, C. Wagner, S. Rubin, J. Goldstein, A. Huber-Lee, C. Young, G. Characklis, and B. Kirsch. 2006. Regional Solutions to Water Supply Provision. Denver, Colo.: Awwa Research Foundation.

Rusk, D. 2003. Little Boxes—Limited Horizons. A Study of Fragmented Local Government in Pennsylvania: Its Scope, Consequences, and Reforms. Washington, D.C.: The Brookings Institution Center on Urban and Metropolitan Policy.

Shih, J.-S., W. Harrington, W.A. Pizer, and K. Gillingham. 2006. Economies of Scale in Community Water Systems. *JAWWA* 98(9):100–108. September.

ABBREVIATIONS

AwwaRF	Awwa Research Foundation
CD	compact disk
GIS	Geographic Information Systems
LVPC	Lehigh Valley Planning Commission
PAC	project advisory committee
U.S.	United States
USEPA	U.S. Environmental Protection Agency